W9-DFH-686

The Tomato in America

Still Life with Vegetables and Fruit by Raphaelle Peale, circa 1795. (Wadsworth Atheneum, Hartford, Ella Gallup Sumner and Mary Catlin Sumner Collection Fund.)

The Tomato in America

Early History, Culture, and Cookery

Andrew F. Smith

University of South Carolina Press

The following recipes are reproduced by permission: "To Make Tomatoe & Ochre Soup," p. 165, from Manuscript cookbook, Mrs. George Read, New Castle, 1813, Holcomb Collection, Historical Society of Delaware, Wilmington; "To Keep Tomatoos for Winter Use," p.181, from Harriott Pinckney Horry Papers, 28, from the Collections of the South Carolina Historical Society.

Published in Columbia, South Carolina, by the
University of South Carolina Press

Manufactured in the United States of America

Library of Congress Cataloging-in-Publication Data

Smith, Andrew F.
The tomato in America : early history, culture, and cookery / Andrew F. Smith.
p. cm.
Includes bibliographical references and index.
Contents: The historical tomato—Historical recipes.
ISBN 1–57003–000–6
1. Cookery (Tomatoes) 2. Tomatoes—History. I. Title
TX803.T6S65 1994 94–3208
641.3'5642'0973—dc20

For the loves of my life:
Tanya, Kelly, and Tim

Contents

Preface

Today, tomatoes are among the most popular fruits or vegetables in the world. From the Americas to Australasia, from northern Europe to southern Africa, the tomato tickles the taste buds of the poor and the rich alike. Globally, more than one and a half billion tons of tomatoes are produced commercially every year. Many more are produced in home gardens. The United States is the largest commercial producer of tomatoes, harvesting about sixteen percent of the world's total. In addition, somewhere between twenty-five and forty million Americans grow tomatoes in their gardens.

Americans devour more than twelve million tons of tomatoes annually, averaging per person about eighteen pounds of fresh tomatoes and almost seventy pounds in processed forms. In addition, America imports about four hundred thousand tons of tomatoes, mainly from Mexico. The value to American farmers of the tomato production is more than one and a half billion dollars annually. The retain market for fresh tomatoes averages about five billion dollars, and additional billions are generated through transporting and processing them. Almost everyone consumes them in salads, soups, juices, ketchups, sauces, salsas, and a feast of other ways.

Tomatoes have not always been so favored. Pop historians report that our forebears eschewed them, considering them poisonous. Others gleefully assert that colonial Americans deemed them aphrodisiacs, called *love apples,* which our Pilgrim forefathers forsook religiously. Some historical gardeners claim that tomatoes were not grown in their region until well after 1860. Agricultural historians proclaim that the tomato did not become a field crop until decades after the Civil War. Culinary historians profess that Americans did not devour them until the mid-nineteenth century, and even then, the tomatoes always were cooked for at least three hours to eradicate their alleged poisonous qualities. Perhaps the most common tomato myth is that of the introduction story, of which there are many versions, reporting what person or group of people first introduced the tomato into America. These myths have one aspect in common: little primary-source evidence has ever been offered in their support. Nevertheless, they have often been recounted in otherwise respectable horticultural, historical, and culinary works.

In exploring these myths and tracing the tomato's tenure in what is today the United States, I have done extensive research in libraries, historical societies, and museums. In the process, I have uncovered more than twelve thousand references to the tomato that were published or written in America before the Civil War. These fell into six major categories: first, botanical, agricultural, and horticultural works, seed catalogues, and gardeners' calendars; second, cookbooks and many cookery manuscripts; third, medical works and periodicals; fourth, newspapers; fifth, menus and bills of fare from restaurants and hotels; and sixth, other works, including travel logs, magazines, other periodicals, almanacs, encyclopedias, dictionaries, regional histories, broadsides, letters, manuscripts, diaries, advertisements, canning labels, reminiscences, autobiographies, and so on. In addition, I have found non-American resources related to the tomato's presence in Mexico, the Caribbean, and Europe for chapter 2, "The Roots of the American Tomato."

Despite the unexpectedly large quantity of material I unearthed, the search has been by no means exhaustive. I have examined only a small fraction of the materials published or written prior to the Civil War. I have not found works believed to have been published; many may not have survived. Of those sources I have examined, I've made no attempt to read each work, although I have skimmed many. In the examination I relied heavily on indexes, in which, undoubtedly, many references to the tomato did not appear. While I located many letters, diaries, and other manuscripts that mentioned tomatoes, I found it impossible to do more than scratch the surface of these voluminous unpublished materials. Likewise, tens of thousands of newspapers were published prior to the Civil War. Inspecting all of them was beyond my scope for this book, though I did examine a sample of approximately five thousand newspapers from all regions of the nation. The large number of tomato references in newspapers indicates the important roles newspapers performed and suggests that a more intensive future examination of them might lead to additional significant findings.

In spite of these qualifications, this book represents the most comprehensive historical study of the tomato in America ever conducted. The history that unfolds is even more bizarre and inherently entertaining than the myths, lore, and pablum frequently regurgitated by newspaper and magazine columnists. The tomato was consumed and cultivated by some Americans during the eighteenth century in all regions of the country, including the South, the Midwest, New England, California, and the American Southwest. By the first decade of the nineteenth century, fresh tomatoes were sold by market gardeners in several states. During the 1820s the adoption of the tomato as a culinary product increased throughout the nation. By the 1830s it was fully integrated into American cookery. During the late 1830s and the 1840s, "tomato mania" struck America. From Maine to California, the tomato was cultivated and consumed. The first part of this book explores American tomato culture and cookery prior to the Civil War, with a closing chapter sketching some more modern tomato phenomena. The second part offers historical recipes for different culinary uses of the tomato. The third includes bibliographic and other resources of interest to tomato lovers.

This book could not have been completed without the assistance of many other researchers, librarians, historians, and friends. In part, the book builds upon the intellectual foundations laid by others. Of particular importance were the works of E. Lewis Sturtevant, George A. McCue, and Karen Hess. *Sturtevant's Edible Plants of the World,* edited by U. P. Hedrick, was a classic attempt to trace the origins of hundreds of plants. Sturtevant's collation of sources on the tomato was particularly helpful. McCue used it as the starting point for his extensive annotated bibliography, "The History of the Use of the Tomato," published in the *Annals of the Missouri Botanical Garden.* He annotated hundreds of citations on the tomato; his bibliography is particularly strong in its European sections. Karen Hess's several works, especially her historical notes and commentaries in the facsimile edition of Mary Randolph's *The Virginia House-wife,* were exceptionally valuable. Her research, findings, opinions, and assumptions were in many ways the starting points for this work, and my debts to her are innumerable. Most of all, I have appreciated her assistance with the cookery and recipe sections and her ongoing encouragement for more than five years. Special thanks also to the late Charles van Ravenswaay, who helped in the initial stages of chapter 4, "Early Tomato Cultivation"; the medical historian J. Worth Estes, M.D., School of Medicine, Boston University, for his useful advice, particularly in reviewing chapters 2 and 6; and Dr. William McNeill, professor emeritus, University of Chicago, for his comments on the European roots of the tomato in chapter 2.

In the process of researching and writing this book, I have tapped materials in the collections of more than four hundred libraries, museums, and historical societies in forty-two states, Australia, Canada, and the United Kingdom. I have made extensive use of the resources at the Brooklyn Botanic Garden, New York Public Library, New-York Historical Society, New York Botanic Garden, and New York Academy of Medicine in New York City; the American Antiquarian Society in Worcester, Massachusetts, and the Schlesinger Library at Radcliffe College in Cambridge, Massachusetts; the Missouri Botanic Garden in St. Louis; the Library of Congress in Washington, D.C.; National Agricultural Library in Beltsville, Maryland, and the National Library of Medicine in Bethesda, Maryland; and the British Library in London.

Many researchers helped locate specific references noted in this book. Several people unearthed significant sources or were otherwise particularly helpful. Among those deserving special mention are Dr. Charles Rick, emeritus professor of vegetable crops at the University of California, Davis; Brother Timothy Arthur, O.F.M., Archivist, Old Mission, Santa Barbara, California; Sibylle Zemitis, reference librarian, California State Library, Sacramento; Constance J. Cooper, library assistant, The Historical Society of Delaware, Wilmington; Rob Blount, Florida Department of Agriculture and Consumer Services, Tallahassee; Marilyn L. Reppun, head librarian, Hawaiian Mission Children's Society Library, Honolulu, Hawaii; John Vander Velde, Farrel Library, Kansas State University, Manhattan; Emelie Willkomm, librarian, the Hermann-Grima Historic House, New Orleans, Louisiana; Jan Longone, Wine and Food Library, Ann Arbor, Michigan; Edward G. Voss, University of Michigan Herbarium, Ann Arbor; Mathew E. Thomas, president, New England Historical Research Associates, Fremont, New Hampshire; Jane R. Odenweller, archivist, Macculloch Hall Historical Museum, Morristown, New Jersey; Marc Simmons, Cerrillos, New Mexico; Robert F. Becker, professor emeritus, Cornell University, New York; Kay Moss, Schiele Museum of Natural History, Gastonia, North Carolina; Kay Bergey, Old Salem Inc., Winston-Salem, North Carolina; Clarissa Dillon, Haverford, Pennsylvania; William Woys Weaver, Paoli, Pennsylvania; John Martin Taylor from Hoppin' Johns Cookbooks, Charleston, South Carolina; Pat Hash, South Carolina Historical Society, Charleston; Pamela Puryear, Navasota, Texas; Phoebe Llyod, Lynchburg, Virginia; Jeff McCormack, Southern Exposure Seed Exchange, North Garden, Virginia; Randall S. Gooden, assistant curator, West Virginia and Regional History Collection, Morgantown, West Virginia; and Ray Swick, Historian, Blennerhassett Historical State Park, Parkersburg, West Virginia. Special thanks also to Alan Fusonie at the National Agricultural Library, who helped locate horticultural and gardening materials; Dennis Lowery at the American Antiquarian Society for his constant help with newspapers; and the staff at the New-York Historical Society library, who frequently indulged my eccentric research interests.

While all suggestions, research, and critical comments were greatly appreciated, this work reflects only my opinions, and I am solely accountable for mistakes that may arise.

The Tomato in America

Part I

THE HISTORICAL TOMATO

Still life with tomato, artist unknown, circa 1840. *(Collection of Mr. and Mrs. Guy M. Mankin.)*

1

Introducing the Introduction Story

On Sunday, January 30, 1949, CBS broadcast live over national radio a reenactment of Robert Gibbon Johnson eating the first tomato in America. This episode of the *You Are There* series depicted an event purportedly held in September 1820 in Salem, New Jersey. According to the CBS broadcast, prior to this date Americans considered the tomato poisonous. Johnson, one of Salem's most prominent citizens, had imported tomato seeds from South America and planted them in his garden. When they produced fruit-bearing vines, he announced that he intended to eat a tomato on the courthouse steps. From hundreds of miles around, spectators traveled to view the sensation. An Italian witness even journeyed all the way from Salem, Massachusetts. (As there were no railroads in southern New Jersey in 1820 and steamboats were still novelties along the Delaware River, a newspaper columnist reviewing the broadcast commented dryly, "The stage coaches, saddle horses, sloops, and schooners must have done a landslide business."[1])

On the appointed day, as the CBS version had it, hundreds of onlookers gathered to see the spectacle of Johnson eating a tomato, expecting him to fall frothing to the ground, then die a painful death. Not all in the crowd, however, believed Johnson would necessarily succumb. The actor playing the part of Dr. James Van Meter, Johnson's personal physician, declared in an interview with the *You Are There* reporter that, while ordinary people would be poisoned by tomatoes, Johnson might not because he had an unusually strong constitution. In the radio version, Johnson mounted the courthouse steps and turned in scorn on the throng. "What are you afraid of ? Being poisoned?" he asked. "Well, I'm not, and I'll show you fools that these things are good to eat." He then sank his teeth into one of the supposedly lethal fruits with dripping relish. Some actor-onlookers fainted. Others gaped in astonishment. But much to almost everyone's surprise, *You Are There* reported, Johnson survived and launched a new and mammoth tomato industry.[2]

The CBS broadcast was not the first rendition of this story. The first known version appeared in 1908, when William Chew, the future publisher of the *Salem Standard & Jerseyman,* asserted simply that Johnson brought tomatoes to Salem in 1820. Sixteen years later Alfred Heston added that after Johnson introduced tomatoes, Salemites considered them inedible but admired them for their appearance. Joseph S. Sickler's initial version of the story, published in 1937, was not much different from Chew's and Heston's statements, but he altered the status of the tomato from an ornamental to an edible plant. Sickler, an amateur local historian, went one step further, claiming that after Johnson introduced the tomato, he patiently educated the natives as to its qualities, showing that it was edible and nutritious. Perhaps he misread Heston or alternatively assumed that, if Johnson introduced the tomato, he probably ate it and persuaded others to do so. Sickler's hypothesis, however speculative, had some basis. Johnson owned a book, published in 1812, that contained a recipe for making tomato ketchup; he therefore knew at least that the cooked tomato was edible.[3]

The Johnson story might have remained simply a part of local lore except that Sickler was also an excellent publicist. He recounted the story to Harry Emerson Wildes, who used it in his nationally acclaimed book *The Delaware.* As Johnson participated in the activities of the local agricultural society and organized fairs in Salem County, Wildes assumed that he had used the agricultural society to push his tomato hobby and had started county fairs to make the vegetable popular. Wildes also claimed that Johnson dared to eat a prize tomato publicly on the courthouse steps. Stewart Holbrook, in his *Lost Men of American History,* dramatized Wildes's version by creating imaginary dialogue for the event.[4]

Joseph Sickler's second version of the story, written in 1948, was not published until the following year. Sickler tried his hand at rewriting the legend, expanding Wildes's and Holbrook's embellishments. His second version differed considerably from his initial brief account. It was enhanced even more dramatically than previous renditions.[5] Sickler then acted as historical consultant to the CBS broadcast and added still more material to the story.

Versions of this story have appeared in professional and scholarly journals and magazines such as *Scientific American, Horticulture,* and *New Yorker;* several publications of historical societies; and prestigious newspapers, including the *New York Times.* Dozens of cookbooks and food books have retold the story. As culinary historian and bookseller Jan Longone has reported, it "has been repeated scores of times in almost yearly articles in major American newspapers and journals." Since 1987 Salem residents have held an annual Robert Gibbon Johnson Day in which a reenactment of the tomato-eating legend is performed on the courthouse steps. The purpose of the reenactment is not to promote historical truth but to foster a sense of pride in Salem's history, which is a laudable goal. Less commendable are those who have reported on the event. In 1988 the reenactment was covered by a Philadelphia television station and rebroadcast over ABC's *Good Morning America.* In this broadcast it was declared that Johnson ate the first tomato in America.[6]

Few writers, journalists, broadcasters, or historians attempted to examine the historical basis for the Johnson story. Those who did performed inadequately. CBS researchers attempted to authenticate it. William Chew affirmed that "Salemites in general seemed to have heard it, but couldn't say just where."[7] This vague response did not deter the network from broadcasting the drama. Francis Coulter wrote to the Salem County Historical Society requesting information about the incident. He received a reply stating that there was no factual evidence that it happened. This response did not stop him from reiterating the story in *Horticulture.* Robert Hendrickson claimed in his book *Lewd Food* that several unnamed scholarly books on the Garden State endorsed the story, but he cited only Coulter's article in *Horticulture.* He made an effort to get at primary sources, he declared, but said it was thwarted when he learned "that a fire in the local newspaper office had destroyed all records of the period." There was no fire in the local newspaper office, but some Salem newspapers from 1820 were missing. Hendrickson stated that the event occurred on September 26, 1820. Unfortunately for his version of the story, a copy of the *Salem Messenger* dated September 27, 1820, survived, as did several subsequent issues. These made no reference to any tomato-eating incident. In fact, no reference to the tomato in Salem has yet been located in any primary source prior to the mid-1830s.

In 1949 Sickler revealed that his version of the story was based on an account given to him by William Casper, a respected auctioneer, many years earlier. According to Sickler, Thomas J. Casper witnessed the incident in 1820 and told the story to his grandson William. The *Salem Standard & Jerseyman* judged that he had a reputation for accuracy in his historic yarns.[8] However, there are serious difficulties with Sickler's attribution of the Johnson story to William Casper. These difficulties are related to William Casper's uncle, Charles Casper.

Charles Casper was a tinsmith who began canning fruit in Salem in 1853. Later he was part-owner of a successful Salem canning factory. He continued to be involved in the canning and packing industry until his retirement at the turn of the century. According to the *Salem Standard &*

Jerseyman, Casper had a good mind and often made valuable contributions to the history of the community. One example was a four-page paper prepared for the Salem County Historical Society on the history of canning in Salem. In this paper he said that "the tomato was brought to Salem, New Jersey, in 1829 by some ladies from Philadelphia." Charles Casper's version of how the tomato arrived in Salem was written in 1906. Charles Casper's father was Thomas Casper, the same person who, according to Sickler, told William Casper the story about Johnson and the tomato. Why would Thomas Casper tell his grandson the Johnson story but not tell his son? As Charles Casper was involved in the canning industry in Salem, he surely would have been familiar with commonly told stories about the tomato, yet he made no mention of the Johnson story in his history of canning.[9]

Salem's history is fairly well documented, particularly during the early nineteenth century. Johnson was often mentioned in Salem's newspapers. His biography appeared in several nineteenth-century works, and his granddaughter compiled a genealogy of the Johnson family. He was an active member of the Salem County Agricultural Society. He also wrote the first history of Salem, which chronologically covered the period around 1820. Despite this large reservoir of information about Salem and Johnson, no nineteenth-century record has yet surfaced crediting Robert Gibbon Johnson with introducing the tomato to Salem.

The Johnson story could be dismissed as simply a good old yarn that accidently became a national legend, but it is not the only tomato introduction story. More than five hundred versions of such accounts have been located, and probably many more are buried in local libraries, newspaper morgues, and historical societies. The majority of these stories have espoused the great-man approach, crediting a particular individual, usually a male, with introducing the tomato into America or to a particular region. The earliest known attribution, published in 1825, purportedly originated with Thomas Jefferson. According to J. Augustin Smith, president of William and Mary College, Jefferson credited London-born Dr. Sequeyra with introducing the tomato into America during the mid-eighteenth century. Hundreds of such attributions have followed regularly. In 1845 Elisha Whittlesey, an ex-congressman from Ohio, expressed the belief that Sardinian-born Francis Vigo introduced the Old Northwest to the tomato in the late eighteenth century. In 1873 Mrs. M. H. Ramsey of Montgomery County, Ohio, stated that "old Mr. Bolton" claimed he introduced tomatoes into this country in the early nineteenth century, having brought the seeds in his pocket from South America and "distributed them among his friends in Philadelphia, who cultivated them for the singularity and beauty of their appearance, but rejected them as an esculent, regarding them as poisonous when grown in this climate."[10]

In 1879 Margery Deane, one of the leaders of high society in Newport, Rhode Island, averred that an immigrant painter from Naples, Michel Felice Corné, prided himself on having set the fashion of eating tomatoes in the 1820s. He reportedly stated

> There . . . is that potato; he grows in the dark or in the damp cellar, with his pale lank roots; he has no flavor; he lives underground. But the tomato, he grows in sunshine; he has fine, rosy color, an exquisite flavor; he is wholesome, and when he is put in the soup you relish him.[11]

In 1900 W. Asbury Christian credited Thomas Jefferson with eating the first tomato in Lynchburg, Virginia. Jefferson often stopped in Lynchburg while going to and from his summer home in Poplar Forest. In 1819 he passed Mrs. Owen's home and, spying tomatoes growing in the yard, asked a girl standing near the gate why she did not eat them. "'Because they are poison,' she replied. 'Bring me one,' said Mr. Jefferson, 'and I will eat it.' She gave him one, and, to her great surprise, he ate it." This story has been regularly improved upon throughout the twentieth century.[12]

In 1903 Mary E. Cutler divulged at a meeting of the Massachusetts Horticultural Society that a man from Bermuda planted the first tomato seeds in America in 1802 in a Pennsylvania jail garden. The prisoner was discharged before they produced bearing vines but returned later to eat their fruit. In 1936 New Jersey's *Bridgeton Evening News* revealed that John Loper grew the first tomatoes in Cumberland County in 1812 from seeds acquired in New Orleans. The organizers of the California agricultural fair decreed that Charles Dendril introduced the tomato into New England in 1837. In 1944 Kenneth Roberts maintained that Yankee sailors brought tomatoes back to Maine well before they were used in the rest of New England. In 1949 newspaper book reviewer Lewis S. Gannett stated that an Italian immigrant, Philip Mazzei, who in 1774 lived on his farm near Thomas Jefferson's Monticello, introduced the tomato to Jefferson. Gannett also reported that the Shakers in Mount Lebanon, New York, believed that the first tomato tasted in America was eaten by a bride who bit into one and refused her doctor-husband's advice to take an emetic, but she had no ill effects. In 1954 the *Cheraw Chronicle* related that Dr. Thomas Powe brought the first tomato plants into South Carolina from Philadelphia in 1822.[13]

In 1967 Griff Niblack, a columnist for the *Indianapolis News,* claimed that Southmayd Guernsey from Clark County, Indiana, purportedly took the historic first bite of a tomato. His grandfather, Daniel Guernsey, who lived in Memphis, had imported some seeds from England and raised the first tomatoes in Indiana, but he thought they were poisonous. Southmayd became a river pilot and visited New Orleans, where tomatoes were served in restaurants. He brought the startling knowledge back to Memphis. He spread the word that he was going to eat a tomato, and presumably it traveled through the town. In an event that suspiciously appears like another Robert Gibbon Johnson incident, a crowd gathered to watch him die. However, Southmayd survived, the neighbors cast aside their doubts, and everybody began to eat tomatoes.[14]

In addition to the great-man approach, cultural groups have been honored as the first to introduce tomatoes into America. In 1839 the *Jeffersonville Courier* claimed that French Huguenot refugees brought them. Forty years later a commentator for the *Mobile Daily Register* speculated that the French colonists grew them along Alabama's Gulf Coast in the early seventeenth century. In 1940 the historian Richard Cummings believed that aristocratic refugees from France introduced them into America after the French Revolution. In 1971 cookbook author Helen Mendes theorized that West Africans cooked them and that slaves who survived the voyage to America were able to do so because they had been fortified with the tomato's vitamins. Others have maintained that the Spanish brought the first tomato to what is today the United States.[15]

Many introduction stories were grounded in fact and have collaborative support, such as contemporary notations in diaries, letters, garden account books, and newspaper reports with eyewitness testimony. Other stories are modern speculations with excellent hypotheses, which, if pursued, might shed insight into the introduction of the tomato into America. Still others are the fanciful inventions of their authors. Some authors, such as those who helped to establish and perpetuate the legend of Robert Gibbon Johnson, viewed the lack of primary evidence as a license to display their creative writing talents. The criterion for selecting (or manufacturing) details was not historical accuracy but good storytelling. While these stories may have made good reading, they made bad history.

The significance of the accounts lies less in their individual factual basis and more in the common questions they addressed or generated. Who introduced the tomato into America? When was it adopted into our cookery? What was the significance of its adoption? For more than 150 years Americans have asked these questions. Despite extensive interest and considerable speculation, they have remained largely unanswered.

NOTES

1. *Salem Sunbeam,* February 1, 1949; *Salem Standard & Jerseyman,* March 12, 1953. The transcript for the broadcast, if one ever existed, has not been located. This rendition is a composite of versions reported in newspapers and other sources. For more information about the legend, see Andrew F. Smith, "The Making of the Legend of Robert Gibbon Johnson and the Tomato," *New Jersey History,* 108 (Fall–Winter 1990): 59–74.

2. Joseph S. Sickler, *The Old Houses of Salem County* (Salem, N.J.: Sunbeam Publishing, 1949), 40; Steward H. Holbrook, *Lost Men of American History* (London: Macmillan, 1946), 131; *Salem Sunbeam,* February 1, 1949.

3. Alfred M. Heston, ed., *South Jersey: A History 1664–1924* (New York: Lewis Historical Publishing, 1924), vol. 1, 487; William Chew, *Salem County Hand Book* (Salem, N.J.: Salem National Banking, 1908), 12; Sickler, *The History of Salem County New Jersey* (Salem, N.J.: Sunbeam Publishing, 1937), 197; James Mease, *Archives of Useful Knowledge* (Philadelphia: David Hogan, 1812), vol. 2, 306.

4. Harry Emerson Wildes, *The Delaware* (New York: Farrar and Rinehart, 1940), 275; Holbrook, *Lost Men,* 131.

5. Sickler, *The Old Houses,* 40.

6. Jan Longone, "From the Kitchen," *American Magazine,* 3 (Autumn-Winter 1987–88): 1; *Salem Sunbeam,* July 22, 1988; July 24, 1988.

7. *Salem Sunbeam,* January 19, 1949.

8. *Salem Sunbeam,* February 1, 1949.

9. Charles W. Casper, untitled report on canning in Salem County. Unpublished manuscript at the Salem County Historical Society, dated March 13, 1906. Walter Hall , the president of the Salem County Historical Society, located it in 1949 and gave it to William Chew, who published parts of it in the *Salem Standard & Jerseyman,* March 24, 1949.

10. Thomas Sewall, *A Lecture Delivered at the Opening of the Medical Department of the Columbian College, in the District of Columbia* (Washington: Printed at the Columbian Office, March 30, 1825), 61; *Western Reserve Magazine of Agriculture and Horticulture,* 1 (July 1845): 100; *Proceedings of the Montgomery County Horticultural Society,* September 13, 1873, pp. 49–51; *Cultivator & Country Gentleman,* 38 (December 11, 1873): 790.

11. *Boston Evening Transcript,* January 20, 1879; George C. Mason, *Reminiscences of Newport* (Newport: Charles E. Hammett, 1884), 339; *Canco,* April 1939, p. 6; Helen Nerney, "An Italian Painter Comes to Rhode Island," *Rhode Island History*, 1 (July 1943): 65–71; Donald J. Boisvert, "Michel Felice Corné: The First Person in America to Eat a Tomato," *Old Rhode Island,* 4 (June-July 1993): 13–16.

12. W. Asbury Christian, *Lynchburg and its People* (Lynchburg, Va.: J. P. Bell, 1900), 84; *(Lynchburg) News,* October 11, 1936; William McKelway, "Jefferson Snack Made Lynchburg Home Famous," *Richmond Times-Dispatch,* August 6, 1978; Dorothy T. Potter and Clifton W. Potter, *Lynchburg: "The Most Interesting Spot"* (Lynchburg, Va.: Beric Press, 1985), 44–45.

13. Mary E. Cutler, "Remunerative Outdoor Occupations for Women," *Transactions of the Massachusetts Horticultural Society for the Year 1903,* part 1 (January 10, 1903): 31–32; Lewis Gannett, *Cream Hill; Discoveries of a Weekend Countryman* (New York: Viking, 1949), 63–65; Kenneth Roberts, *Trending Into Maine* (Garden City, N.Y.: Doubleday, Doran, 1944), 147; *Bridgeton Evening News,* May 24, 1936; *Cheraw Chronicle,* March 25, 1954; W. Armstrong Price to Lewis S. Gannett, dated October 11, 1940, Virginia Historical Society, MS52 P93175b.

14. *Indianapolis News,* September 19, 1967.

15. *Jeffersonville Courier,* as in the *Connecticut Courant,* June 15, 1839; *Mobile Daily Register,* June 1, 1879; Richard Osborn Cummings, *The American and His Food: A History of Food Habits in the United States* (Chicago: University of Chicago Press, 1940), 31; Helen Mendes, *The African Heritage Cookbook* (New York: Macmillan, 1971), 36, 55.

2

The Roots of the American Tomato

NAMES AND ALIASES

The sixteenth century was a period of scientific ferment in Europe, particularly in the field of botany. Overwhelmed as new plants arrived simultaneously from Africa, Asia, and the Americas, European botanists classified them into contexts they understood. In the mid-sixteenth century these contexts encompassed religious beliefs, resuscitated ancient Greek and Roman myths, rudimentary botanical classification systems, and emerging medical theories. Herbalists, botanists, and physicians compiled their observations and beliefs in herbals composed of descriptions of plants and their purported medicinal virtues. They assigned names to the new plants as they thought appropriate, but it quickly became apparent that they had named the same plant differently. Likewise, similar names were assigned to decidedly different plants, resulting in bewilderingly long lists of conflicting names.

Renaissance herbals were modeled after ancient Greek and Roman manuscripts attributed to Dioscorides and Galen. Dioscorides, a Greek doctor in the Roman army, traveled widely and knew the plants of the Mediterranean. His only surviving work, *Materia Medica,* was completed about the middle of the first century. In part his renown can be attributed to the praise offered by Galen, a Greek physician living a century later in Pergamum and Rome. Many of Galen's works survived. He emphasized a medical system based on the notion of humors, qualities, and spirits. According to this system, there were four humors in the human body: black bile, yellow bile, blood, and phlegm. Humors had four qualities: two active (hot and cold) and two passive (wet and dry). The human body also had vegetative, vital, and animal spirits. Galen believed that food was converted into blood in the liver, the vegetative part of the body. The heart, the seat of passion, converted the blood into vital spirits. The brain, the seat of the rational and immortal qualities, added the animal spirits.

When these humors, qualities, and spirits were out of balance, illness resulted. The basic goal of Galenic medicine was to reestablish balance through purifying humors and evacuating defective ones. In herbals, the medicinal effects of plants were associated with particular humors. These effects were measured by degrees, which were noticeable differences in the strength of medicinal substances in the plant. Degrees ranged from temperate (neutral) through mild (first degree) to the strongest possible (fourth degree), which were usually poisonous or caustic. Galen's beliefs were revitalized during the Renaissance. As will be demonstrated in a later chapter on the medicinal virtues of the tomato, these beliefs survived to the mid-nineteenth century.

Using the works and principles of Dioscorides and Galen as their guides, Renaissance herbalists tried to locate each plant mentioned in the ancient manuscripts to rediscover its medicinal virtues as claimed by the classical Greeks and Romans. They initially assumed that Galen, Dioscorides, or other physicians had located and catalogued all existing plants, and early herbals were commentaries on the ancient manuscripts. As herbalists began collecting and describing plants then growing in Europe, however, it soon became apparent that many had not been mentioned by the classical Greeks. New systems of classifying plants needed to be developed.

In 1544 an Italian herbalist, Pietro Andrae Matthioli, published a reference to *mala aurea,* or "golden apples," which he described as "flattened like the melrose [sort of apple] and segmented, green at first and when ripe of a golden color."[1] This was the first known European reference to the tomato. It suggests that the tomatoes conveyed initially into Europe were yellow in color. Matthioli classified them with the mandrake plant. In turn, mandrake was classified along with the nightshades, many of which were toxic. Perhaps mandrake acquired its undeserved toxic reputation through association with these plants. Its real claim to fame rested with other purported qualities. It had been mentioned in the Bible. The Hebrew term for it was *dudaïm,* which translated into English as "love apples" or "love plants." In Genesis, Rachel and Leah employed mandrake roots as a love potion.

Several ancient civilizations believed in the mandrake's aphrodisiac powers. This belief derived from the "doctrine of signatures," which held that the medical benefits of a plant could be determined from its external appearance. The Greeks believed that mandrake roots looked like two entwined lovers and consequently considered them to have aphrodisiac qualities. The doctrine of signatures survived into the Middle Ages and became fashionable during the Renaissance.[2]

Matthioli recounted that the golden apples were cooked in the same way as eggplants: fried in oil with salt and pepper. The eggplant, a product of South Asia, was an established food product in Spain and Italy by the early sixteenth century. Herbalists classified it as a type of mandrake and referred to it as a love apple, probably because of its botanical similarity to the mandrake. Since mandrake, eggplants, and tomatoes are classified today as members of the *Solanaceae* family, it is no accident that early herbalists classified them together: their leaf structures, flowers, and fruit were similar. Matthioli enlarged his herbal and published a second edition in 1554. This edition gave the Italian name for the *mala aurea* as *pomi d'oro,* or "golden apples," a name that continues to be used in Italy to this day. In this edition Matthioli also mentioned a red variety.[3]

Herbalists linked tomatoes with classical Greek texts. Some concluded that Matthioli's golden apples were similar in appearance to the descriptions of Dioscorides' *glaucium,* which was the juice of a bitter, strong-scented, saffron-colored herb grown in Syria. Dioscorides recommended *glaucium* as a salve for inflammations of the eye. Other herbalists thought that the golden apples were Galen's *lycopersicon* (wolf's peach), a strong-smelling, yellowish liquid that derived from a North African plant. The twentieth-century botanist Leonard Luckwill pointed out that the actual transliteration from Greek is *lycopersion,* a word or phrase of unknown meaning.[4] Whatever Galen may have meant, later herbalists interpreted this to be *lycopersicon.* Some added a Latin ending on the Greek word, resulting in the term *lycopersicum,* which is still frequently used today.

Matthioli's golden apples were neither Dioscorides' *glaucium* nor Galen's *lycopersicon,* but they acquired these names and the properties assigned to them by the ancient herbalists. Renaissance herbalists recommended the juice of the tomato as a potion for cataracts and other inflammations of the eye. Similar medicinal properties were assigned to tomatoes by later herbalists and physicians, and these attributions have remained with them into the twentieth century.[5]

In 1553 the Swiss naturalist Konrad Gesner painted a watercolor of a small, red-fruited plant that he called in Latin *poma amoris* (love apple). This name may have been given to the small-fruited tomato because of its similarity to the mandrake fruit. The attribution may have originated with Luca Ghini, a sixteenth-century Italian botanist who had founded the first European botanical garden at Pisa. He called the fruit *amatula,* a Latin word that denotes possession of an aphrodisiac quality. Ghini corresponded with Matthioli and almost every other European herbalist. He may have been the source for Matthioli's initial information on the *mala aurea.*[6] In any case, a few observers mentioned that tomatoes were considered aphrodisiacs, as were many other new plants introduced into Europe from Asia, Africa, and the New World.

Gesner too painted a picture of a large, lumpy-fruited plant called *Solanum lycopersicum,* which he believed was related to the small-fruited love apple. In 1553 the Flemish herbalist Rembert Dodoens also concluded that Matthioli's golden apples were the same as the *poma amoris.* Dodoens's second edition of his herbal, published the following year, contained a woodcut of the large-fruited variety, which was deeply furrowed, lumpy, and flat at both ends. This herbal was revised, expanded, and translated into French by Charles de l'Écluse, a colleague of Dodoens. L'Écluse cited the classical Greek myth of the Hesperides, who were the guardians of the golden apples given to Hera at her wedding to Zeus. The golden apples were tokens of eternal life and divine fertility. The Hesperides were assisted by a dragon, which was slain by Hercules. Hercules then carried off the apples. The Hesperides myth both resembled and differed from the biblical account in the Garden of Eden. The myth, along with its association with the tomato, was published in England by Henry Lyte, who translated the Dodoens and l'Écluse herbal in 1578, and later the myth was repeated by John Gerard in his *Herball.*[7]

THE TOMATO'S BOTANICAL ORIGIN

Renaissance herbalists also sought to identify the locations where plants originated, though they had a crude geographical understanding of the world. Tomatoes purportedly came from a variety of locations. The terms *glaucium* and *lycopersicon* suggested a Mediterranean origin. Luigi Squalermo, the first prefect of the Padua Botanical Garden, used the term *pomi de Peru,* which implied a South American origin. However, he probably confused the plant with *Datura stramonium,* which was also called *pomi de Peru.* The German herbalist Joachim Camerarius called the plant *pomum Indium,* which may have meant an Asian origin. Guilandinus, the second prefect of the Padua Botanical Garden, used the term *Americanorum Tumatle pai tumatle* to refer to the *pomum aureum* and the *pomum amoris.* As the twentieth-century botanist J. A. Jenkins has pointed out, Guilandinus used the word *Themistitan,* a corruption of Tenochtitlan, which was another name for the present-day Mexico City. As this term was used for only a few years after the Spanish Conquest, Jenkins maintained that the tomato arrived in Europe during the 1520s.[8]

The issue of where the tomato plant originated was not resolved until the mid-twentieth century. The account that emerged was that the tomato

(*Lycopersicon*) originated in the coastal highlands of western South America. Wild tomato plants can still be found in the coastal mountains of Peru, Ecuador, and northern Chile. No evidence has been uncovered indicating that any indigenous South American group cultivated or even ate tomatoes prior to the Spanish Conquest. Through natural means, probably through their ingestion by turtles in South America, wild tomatoes migrated to the Galapagos Islands. Through some unknown means, *Lycopersicon* migrated to Central America. Mayan and other Mesoamerican peoples domesticated the plant and used its fruit in their cookery. The wild tomato was two-celled. A genetic mutation occurred, producing a multi-celled fruit, which was large and lumpy. Central American farmers nurtured and developed this mutation. Today's large, smooth-skinned fruits are mainly crosses between the large, lumpy-skinned mutations and the smooth-skinned cherry tomato.[9]

Unlike many other New World fruits and vegetables, the cultivated tomato was found only in Central America. As no pre-Columbian archaeological evidence of the tomato has been uncovered, no pottery has been located with tomatoes on it, and no hieroglyphics have been found that mention them, most observers have suspected that tomatoes were a late addition to the food supply of the Mesoamericans. The Aztecs readily adopted them, probably because of their similarity to the *tomatl* (*Physalis ixocarpa*)—a plant believed to be native to the Mexican highlands. The *tomatl* fruit was small, green or yellow in appearance, sour in taste, and covered with a paper-thin membrane. The Aztecs named the new plant *xitomatl,* or large *tomatl.*[10] By the sixteenth century this fruit was cultivated at least in the southern part of Mexico. The Spanish first encountered it after Hernán Cortés began his conquest of Mexico in 1519.

The Spanish Tomato

Several Spanish accounts written after the Conquest offered insight into the pre-Columbian culinary use of the *xitomatl.* Joseph de Acosta, a Jesuit priest who lived most of his life in Mexico, stated that *tomates* were "cold and very wholesome" and "full of juice, which gives a good taste to sauce, and they are good to eat." Bernardino Sahagún, a Franciscan priest who went to Mexico in 1529, asserted that Aztecs combined the *xitomatl* with chilies and ground squash seeds to make a sauce (or salsa). This sauce was served with fish, lobster, sardines, turkey, venison, other meats, and seafood.[11]

Despite the similarities in the Aztec names, the *tomatl* and *xitomatl* were only remotely related botanically. Initially Spaniards lumped them together and translated both with the word *tomate,* thus occasionally leaving some confusion in early accounts as to which plant the word *tomate* referred. This confusion still exists in the English language as the terms *husk tomato, green tomato,* or *tomatillo* (*Physalis ixocarpa*) refer to the original Aztec *tomatl* and the term *tomato* (*Lycopersicon esculentum*) to the original *xitomatl.*

As a general practice, the Spanish distributed desirable fruits and vegetables throughout their empire. Although the cherry tomato had originated in Peru and northern Chile, no evidence has been uncovered that indicates that the large, lumpy *tomate* was cultivated or consumed in South America prior to the Spanish Conquest. The Spanish also introduced the tomato into the Caribbean and the Philippines. From the Philippines, its culture dispersed to Southeast Asia and ultimately the rest of Asia. Through the Spanish, the tomato and the husk tomato were also disseminated into Europe. Only the tomato thrived. The reason for this success may have been botanical. The *Physalis ixocarpa* is self-incompatible, which means it will not self-pollinate. In self-pollinating plants, fertilization occurs when the pollen and eggs from the same flower combine. More than ninety-five percent of the seeds found in the modern tomato fruit are the result of self-pollination. Most highly successful domesticated plants, including the tomato, are self-compatible. When a mutation arises in the tomato plant, such as the large, lumpy variation of the small, smooth-skinned cherry tomato, it is easier to perpetuate. Hence the fruit of the tomato plant comes in a wide variety of shapes, sizes, and colors. Because of its great variability, it was perhaps no wonder that early herbalists identified the plant variously as *glaucium, lycopersicon,* golden apples, and love apples.

Tomato plants were not always self-compatible. In fact, many species of wild tomato plants rarely self-pollinate and usually require bees to ensure cross-pollination. The reason for this crucial difference between many wild varieties and modern tomato types requires some explanation. The tomato flower has both the male part, or stamen (composed of the anther and filament), and the female part, or pistil (composed of the stigma, style, and ovary). In many wild tomato types, the stigma extends beyond the anther cone, making it difficult for the plant to self-pollinate. The stigma is accessible to bees, which then carry the pollen to other flowers, producing cross-fertilization. In modern tomatoes the style is shorter than the anther cone. When the tomato flower opens, even slight movements of the plant knock the pollen off the anther onto the stigma, where pollination occurs. The pollen produces a long, thin tube that grows through the stigma and style and into the ovary, where it fertilizes the eggs. For every seed in the tomato fruit, one pollen grain and one egg have combined. When the tomato plant was introduced into Europe, the halictus bees, which pollinated the plant in its natural habitat, were left behind. Plants with shorter styles produced more fruit and therefore had better chances for survival. Farmers may have also encouraged this quality by saving and sowing seeds from the plants whose fruit was more prolific.[12]

Tomatoes grew easily in the Mediterranean climate of Spain and Italy. They were used for culinary purposes in Seville at least by 1608, probably in a salad along with cucumbers. As culinary historian Rudolf Grewe noted, the earliest known cookbook with tomato recipes was published in Naples in 1692. However, the author identified the recipes as Spanish in origin.[13] Additional recipes were published in Italy and later in Spain during the eighteenth century.

The British Tomato

Although tomatoes had been cultivated in Continental Europe since the 1540s, they were not grown in England until the 1590s. John Gerard, a barber-surgeon, planted them in Holborn in the College of Physicians gardens that he superintended. Gerard's *Herball,* initially published in 1597, was plagiarized mainly from the works of Dodoens and l'Écluse. Gerard was so anxious to publish it that many errors appeared in the first edition, such as the misspelling *Lycoperticum.* Thomas Johnson, the editor of the second edition, corrected this mistake as well as countless others without comment.

In one of the original sections of the herbal, Gerard claimed that his golden apples or apples of love came from Spain and Italy. He also stated that the Spanish and Italians ate them "boiled with pepper, salt and oile." Gerard considered "the whole Plant" to be "of ranke and stinking savour." He believed that the temperature of the tomato was "colde, yet not fully so colde as Mandrake, after the opinion of Dodonaeus: but in my judgement it is very colde, yea perhaps in the highest degree of coldeness." The fruit was corrupt, which he left to every man's censure. While the leaves and stalk of the tomato plant are toxic, the fruit is not. Despite the fact that Gerard knew that the tomato was eaten in Spain and Italy, he accepted Dodens's view that it was poisonous. His negative views prevailed in Britain and in the British North American colonies for over two hundred years. John Parkinson, the apothecary to King James I and botanist for King Charles I, proclaimed that, while love apples were eaten by the people in the hot countries to "coole and quench the heate and thirst of the hot stomaches," British gardeners grew them only for curiosity and for the amorous aspect or beauty of the fruit.[14]

In his herbal *Botanologia,* published in 1710, William Salmon alleged that in northern Africa, Egypt, Italy, Spain, and other "hot Countries" tomatoes were boiled with vinegar, pepper, and salt and served with oil and lemon juice. They were also eaten raw "with Oil, Vinegar and Pepper for Sawces to their Meat." In England they were grown in gardens for medicinal purposes. Salmon believed that they had "alterative" qualities that cleansed the system and otherwise improved health. He recommended tomato juice for inflammations; tomato "Essence" for repressing "Vapors in Women" and "Fits of Mother" and for opening obstructions in the bladder; tomato cataplasm (or poultice) for headaches, gout, and sciatica; tomato oil for burns; and tomato balsam for cleansing "Running Sores" and for ulcers, inflammations in wounds, back pains, and many other ailments and diseases. Although Salmon was not highly respected among his professional colleagues and was considered a quack and mountebank by subsequent historians, he was a popular and prolific writer, and his recommendations for the medical use of tomatoes may have influenced others.[15]

Reports of the usage of tomatoes regularly appeared in British publications. In 1719, according to the English edition of Joseph Pitton de Tournefort's *The Compleat Herbal,* the Italians ate them, "but considering their great Moisture and Coldness, the Nourishment they afford must be bad." However, the juice of the tomato was used to cure "Rheum or Defluxion of hot humours upon the Eyes, which may occasion a Glaucoma if not prevented. For it not only stops the afflux of the Humour but moderates and allay the influxion." It was also recommended for St. Anthony's fire (erysipelas). When the fruit was boiled in oil, it was used for curing itches and burns. Tournefort was the first botanist to list *Lycopersicon* as a separate genus within what was then called the *Solanum* family and later was renamed *Solanaceae.* He listed nine species, most of which are today not considered part of the *Lycopersicon* genus.[16]

In 1728 Richard Bradley, a professor of botany at Cambridge University, said that every sort of tomato that he had seen "makes an agreeable Plant to look at, but the Fruit of most of them is dangerous." Gardeners advised planting them away from habitation, "for the leaves and stalks, when rubbed by the clothes in people's passing by, yield a very strong and very offensive smell." Such views were common in England well into the nineteenth century. As late as 1831, the British horticulturist Henry Phillips recounted that the tomato plant had "a rank smell; on which account it was formerly called Malum Aureum odore foetido, the stinking golden apple."[17]

In 1731 Philip Miller, the superintendent of the Chelsea Physick Garden, reported that small yellow love apples were "directed for Medicinal use, by one College in their Dispensatory." Miller continued to grow and observe tomatoes. In 1752 he admitted that they were much used in England for soups. He hastened to add, however, that "there are persons who think them not wholesome." Miller, a friend of Carl von Linne (more commonly known as Linnaeus), employed the latter's binomial classification system. Linnaeus classified the tomato as part of the *Solanum* family, giving its scientific name as *Solanum lycopersicum.* Miller disagreed. In 1754 he classified it as a genus separate from *Solanum,* naming it *Lycopersicon esculentum,* or "edible wolf's peach." Debates as to the tomato's proper botanical classification continued well into the twentieth century, but eventually taxonomists agreed with Miller. His was later recognized as the first valid publication of the tomato's scientific name. However, recent historical studies have questioned this. Dr. James L. Reveal, professor of botany at the University of Maryland, has located an earlier, validly published reference to *Lycopersicum esculentum,* written by John Hill in a supplement to Chambers' *Cyclopedia.*[18]

The specific members of the *Lycopersicon* genus and its precise boundaries with some closely-related plants in the *Solanum* genus have remained unclear. In principle, a genus is composed of homogeneous members. Some members of the *Lycopersicon* genus, such as *L. esculentum, L. cheesmanii,* and *L. pimpinellifolium,* are easily cross-bred and produce many accessions. There are fewer connections among these species and others within the genus, such as *L. chilense, L. hirsutum,* and *L. peruvian.* They can be easily cross-bred with some members of the *Solanum,* however, such as the *Solanum lycopersicoides.* Of the nine generally accepted species of *Lycopersicon,* only *L. esculentum* and, to a much smaller extent, *L. pimpinellifolium* (otherwise called the currant tomato) are used for culinary purposes. The other species produce fruit, but most tasters consider it to be repulsive. Scientific interest in the other species has been generated for purposes of breeding particular traits, such as resistance to certain diseases, into *L. esculentum.* Recent genetic studies of tomatoes, potatoes, and other related plants by David M. Spooner, Gregory Anderson, and Robert Jansen offered evidence suggesting that tomatoes are a part of the *Solanum* genus.[19]

In 1751 John Hill contended that the fruit was eaten by many people, but it ought to be thoroughly ripe. Two years later Hill reported that the tomato was eaten stewed or raw by Jewish families in England. This comes as no surprise. Many Jewish families living in England were engaged in trade with the Caribbean and the Americas. Many were of Portuguese or Spanish descent and had migrated from or maintained contact with Jewish communities in the New World who consumed tomatoes. At least one English-born Jewish physician introduced tomatoes into Virginia during the mid-eighteenth century.[20]

Jewish families were not the only ones to eat tomatoes in England. In 1758 a supplement to Hannah Glasse's *The Art of Cookery* included a recipe,

"To Dress Haddock after the Spanish Way," in which love apples were a seasonal ingredient.[21] John Abercrombie reported that tomatoes were used in some families for improving soups, pickles (both green and ripe), stews, and sauces. By the 1780s tomato sauce was put on anything sent to the table. Other tomato recipes appeared in British cookery manuscripts. Before the end of the eighteenth century the *Encyclopædia Britannica* announced that the tomato was "in daily use; being either boiled in soups or broths, or served up boiled as garnishes to flesh-meats."[22]

THE CARIBBEAN AND THE TOMATO

Many British authors claimed incorrectly that *tomato* was a Spanish or Portuguese word. While it was a corruption of the Spanish word *tomate,* which in turn derived from the Aztec *tomatl,* the word *tomato* probably originated in Jamaica. The Irish-born Hans Sloane, a pupil of Joseph Pitton de Tournefort and one of the foremost naturalists of his age, botanized in British-controlled Jamaica late in the seventeenth century. He found the *tomato berry* growing wild around St. Iago de la Vega, previously the capital of Spanish Jamaica. As the commonly used English, French, and German words in the late seventeenth century were the equivalents of *love apple,* the word *tomato* may have originated with the Spanish slaves who were not evacuated when the British seized the island. Whatever its origin, the word was promptly picked up in England and in the British North American colonies. According to the botanist William Hanbury, the term *tomato* had become fashionable by 1771. It did not achieve its predominant linguistic position in the English language until the nineteenth century, however.[23]

Tomatoes were eaten in Jamaica and probably throughout the Caribbean. In the early eighteenth century Henry Barham confessed that he had eaten five or six raw tomatoes at a time in Jamaica. They were "full of pulpy juice, and of small seeds, which you swallow with the pulp, and have something of a gravy taste." Barham believed that the juice of the tomato was cooling "and very proper for defluxions of hot humours in the eyes, which may occasion a glaucoma, if not prevented; [tomatoes] are also good in the St. Anthony's fire, and all inflammations; and a cataplasm of them is very proper for burns." Patrick Browne mentioned that tomatoes were roasted and used with mutton. Edward Long said that the juice of the tomato was "often used in soups and sauces, and imparted "a very grateful flavour." Tomatoes were likewise fried, and served with eggs. Long also disclosed that the "Spaniards esteem them aphrodisiacs."[24] This purported quality was rarely mentioned in herbals or other sources. Despite the name love apples, which continued to be used in the English language until the mid-nineteenth century, there was little evidence that anyone in America or Britain ever considered the tomato to be an aphrodisiac.

NOTES

1. Pietro Andrae Matthioli, *Di pedanio Dioscoride anazarbeo libri cinque della historia,* translated by George A. McCue, "The History of the Use of the Tomato: An Annotated Bibliography," *Annals of the Missouri Botanical Garden,* 39 (November 1952), 292.

2. The doctrine of signatures was in common usage by the Greeks and Romans and was in vogue at the time of Pliny. Simply stated, it held that natural products, especially plants, had signs on them indicating what kinds of diseases they were useful for treating. For instance, walnuts were purportedly good for head injuries because of their visual similarity to the skull and brain. Arthur H. Graves, "The Doctrine of Signatures," *Herb Grower Magazine,* 7 (January-March 1953): 3–6; Helen Bancroft, "Herbs, Herbals, Herbalists," *Scientific Monthly,* 35 (September 1932): 252; A. G. Morton, *History of Botanical Science* (London: Academic Press, 1981), 224; Agnes Arber, *Herbals, Their Origin and Evolution: A Chapter in the History of Botany 1470–1670* (Cambridge: University Press, 1938), 250–55.

3. Ruperto de Nola, *Libro de cozina cõpuesto* (Toledo: Ramon de Petras, 1525), xxiii; Giovanni Francesco Angelita, *I pomi d'oro di Gio. Francesco Angelita doue si contengono due lettioni de'fichi l'vna, e de'melloni l'altra* (Ricanati: Antonio Braida, 1607); Leonhart Fuchs, *De historia stirpivm* (Baselae: Isingriniana, 1542), 332, 517; Joannes Ruellius, *De natura stirpium libri tres* (Basileae: in off. Frobeniana per H. Frobenium, 1537), 441; Matthioli, *Commentarii in libros sex pedacii Dioscoridis Anazarbei de medica materia* (Venetiis: In officina Erasmiana apud Valgrisium, 1554), 479.

4. Leonard C. Luckwill, *The Genus Lycopersicon: An Historical, Biological and Taxonomic Survey of the Wild and Cultivated Tomatoes,* Aberdeen University Studies No. 120 (Aberdeen: University Press, 1943).

5. Robert T. Gunther, *The Greek Herbal of Dioscorides* (London: Hafner, 1968), 333; Luigi Squalermo [Anguillara], *Semplici dell' eccellente* (Vinegia, V. Valgrisi, 1561), 217; Petrus Pena and Mathia de l'Obel, *Stirpium adversaria nova* (London: Thomas Purfoot, 1570), 106, 108–9; Melchioris Guilandini [Guilandinus], *Papyrus; hoc est commentarius in tria C. Plinii maioris de papyro capita* (Venetiis: apud M. A. Ulmum, 1572), 90; Ioachimus Camerarius, *Hortus Medicus* (Francofurti ad Moenum, 1588), 130; John Gerard, *The*

Herball or Generall Historie of Plantes (London: Printed for the Author, 1597), 275; John Parkinson, *A Garden of Pleasant Flowers; Paradisi in Sole: Paradisus Terrestris* (New York: Dover, 1976), 379–80.

6. Ioh. Bauhin and Ioh. Cherlerus [Dominicus Chabraeus, ed.], *Historia plantarum universalis* (Ebrodumi: 1651), vol. 3, 621; George A. McCue, "The History of the Use of the Tomato: An Annotated Bibliography," *Annals of the Missouri Botanical Garden,* 39 (November 1952): 292.

7. Conradi Gesner, *Historia plantarum,* Faksimileausgabe [Heinrich Zoller, Martin Steinmann, and Karl Schmid, eds.], (Zürich: Urs Graf-Verlag, 1974), vol. 3, figures 7 and 11, which is dated September 22, [15]53; Rembertus Dodonaeus, *Trium priorum de stirpium historia* (Antverpiae: 1553), 428; Rembert Dodoens, *Cruyde-Boeck*, facsimile (Amsterdam: 1971); Charles de l'Écluse, *Histoire des plantes de Rembert Dodoens,* facsimile with an introduction by J. E. Opsomer, (Bruxelles: Centre National D'Histoire des Sciences, 1978); Henry Lyte, *Nievve Herbal or Historie of Plantes* (Antwerp: Henry Loë for Gerard Dewes, London, 1578), 439–40; John Gerard, *The Herball or Generall Historie of Plantes* (London: Printed for the Author, 1597), 275–76.

8. Squalermo, *Semplici,* 217; Asa Gray and J. Hammond Trumbull, "Review of DeCandolle's Origin of Cultivated Plants with Annotations upon Certain American Species," *American Journal of Science,* 26 (August 1883): 128; Camerarius, *Hortvs,* 70; Guilandini, *Papyrus,* 90; J. A. Jenkins, "The Origin of the Cultivated Tomato," *Economic Botany,* 2 (October-December 1948): 379–92.

9. L. A. M. Riley, "Critical Notes on Galapagos Plants," *Bulletin of Miscellaneous Information,* Royal Botanic Gardens, Kew, 5 (1925): 227; Charles M. Rick, "Genetic and Systematic Studies on Accessions of Lycopersicon from the Galapagos Islands," *American Journal of Botany,* 43 (1956): 687–96.

10. Robert L. Dressler, *Botanical Museum Leaflets, Harvard University No. 6,* 16 (December 1953): 137; J. N. Rose, "Notes of Useful Plants of Mexico," *Contributions to the U.S. National Herbarium* (Washington, D.C.: 1899), vol. 5, 210.

11. Joseph de Acosta [Edward Grimston, trans.], *The Naturall and Morall Historie of the East and West Indies* (London: Printed by Val Sims for Edward Blount and William Aspley, 1604), 294; Bernardino de Sahagún, *General History of the Things of New Spain (Florentine Codex),* as in Charles Dibble and Arthur J. O. Anderson, trans., Book 8—Kings and Lords, Monograph Number 14, part 9 (Santa Fe, N.M.: School of American Research and University of Utah, 1961), 37–38.

12. *Tomato Club,* 1 (May 1993): 1–2.

13. Hospital de la Sangre, *Libros de recibos y gastos,* legajo 455, as in Michael Allen and Robert Benson, eds., *First Images of America: The Impact of the New World on the Old* (Berkeley: University of California Press, 1976), vol. 2, 859; Antonio Latini, *Lo Scalco alla Moderna* (Napoli: con Licengadé superiorise privilegio, 1692–94), vol. 1, 390, 444, 551; vol. 2, 55, 162; Rudolf Grewe, "The Arrival of the Tomato in Spain and Italy: Early Recipes," *The Journal of Gastronomy,* 3 (Summer 1987): 67–83.

14. Benjamin Daydon Jackson, ed., *A Catalogue of Plants Cultivated in the Garden of John Gerard in the Years 1596–1599* (London: Privately printed, 1876), 14; Gerard, *The Herball*, 275–76; John Parkinson, *A Garden of Pleasant Flowers; Paradisi in Sole: Paradisus Terrestris,* reprint, (New York: Dover, 1976), 380.

15. William Salmon, *Botanologia. The English Herbal or, History of Plants* (London: Printed by I. Dawks for H. Rhodes and J. Taylor, 1710), 29–30; C. J. S. Thompson, *Quacks of Old London,* reprint, (New York: Barnes & Noble, 1993), 125–33.

16. Joseph Pitton de Tournefort, *The Compleat Herbal: or The Botanical Institutions of Mr. Tournefort* (London: R. Bonwiche, 1719), vol. 1, 214–15.

17. Richard Bradley, *Dictionarium Botanicum: or, a Botanical Dictionary for the Use of the Curious in Husbandry and Gardening* (London: Printed for T. Woodward & J. Peele, 1728), appendix to vol. 2; Robert Gunther, *Early British Botanists and their Gardens Based on the Unpublished Writings of Goodyear, Tradescant, and Others,* reprint, (New York: Kraus Reprint, 1971), 50; Ephraim Chambers, *Cyclopedia* (London: W. Strahan, 1781), vol. 3; Henry Phillips, *The Companion for the Orchard* (London: Henry Colburn and Richard Bentley, 1831), 225–27.

18. James L. Reveal, "Two Previous Unnoticed Sources of Generic Names Published by John Hill in 1773 and 1774–5," *Bulletin du Muséum National de Histoirie Naturelle,* Ser. 4, 13 (1991): 222. In September 1993, the Botanical Congress met in Yokohama and passed a provision to Article 13 of the International Code of Botanical Nomenclature specifically rejecting publications by John Hill. Philip Miller, *Gardeners Dictionary* (London: Printed for the Author, 1731, rev. 1752, 1754).

19. Charles Rick, "Biosystematic Studies in *Lycopersicon* and Closely Related Species of *Solanum,*" in J. G. Hawkes, R. N. Lester, and A. D. Skelding, eds., *The Biology and Taxonomy of the Solanaceae* (London: Academic Press, 1979), 667–77. David M. Spooner, Gregory J. Anderson, and Robert K. Jansen, "Chloroplast DNA Evidence for the Interrelationships of Tomatoes, Potatoes, and Pepinos (Solanaceae)," *Journal of Botany,* 80 (1993): 676–88.

20. John Hill, *A History of Plantes* (London: Thomas Osborne, 1751), 296; *A Supplement to Mr. Chambers's Cyclopedia: or, Universal Dictionary of Arts and Sciences* (London: Printed for W. Innys and J. Richardson, 1753), vol. 2; John G. Stedman, *Narrative of a Five Year Expedition against the Revolted Negroes of Surinam in Guiana on the Wild Coast of South America from the Years 1772–1777,* introduction and notes by R. A. Lier (Holland: Imprint Society, 1971), vol. 2, 343; Phillips, *Pomarium Britannicum: An Historical and Botanical Account of Fruits Known in Great Britain* (London: Printed for the Author, 1820), 236; Thomas Sewall, *A Lecture Delivered at the Opening of the Medical Department of the Columbian College, in the District of Columbia* (Washington: Printed at the Columbian Office, March 30, 1825), 61.

21. Hannah Glasse, *The Art of Cookery Made Plain and Easy* (London: Printed for the Author, 1758), 341. As Glasse's connection with the cookbook had been severed before the publication of this edition, the author of the recipe is unknown. Glasse's cookbook was the most popular one in England and in British North America during the latter part of the eighteenth century.

22. Thomas Mawe [pseud. for John Abercrombie], *Every Man His Own Gardener*, 4th ed. (London: W. Griffin, 1769), 110, and subsequent editions; from a cookery manuscript at Towneley Hall in Lancashire quoted by Elizabeth David, as in John L. Hess and Karen Hess, *The Taste of America* (New York: Grossman, 1977), 81; *Encyclopædia Britannica* (Edinburgh: A. Bell & C. Macfarquhar, 1797), vol. 17, 597–98.

23. *A Supplement to Mr. Chambers's Cyclopedia: or, Universal Dictionary of Arts and Sciences* (London: Printed for W. Innys and J. Richardson, 1753), vol. 2; William Hanbury, *A Complete Body of Planting and Gardening* (London: Printed for the author, 1771), vol. 2, 752.

24. Henry Barham, *Hortus Americanus* (Kingston, Jamaica: Alexander Beekman, 1794), 92–93; Patrick Browne, *History of Jamaica* (London: Sold by B. White and Son, 1789), 175; Edward Long, *The History of Jamaica* (London: Frank Cass and Co., 1970), 47.

3

The Arrival of the Tomato in America

Besides selling medicines and writing books, the English herbalist William Salmon was an outspoken critic of Catholicism. In 1687, during the pro-Catholic reign of James II, Salmon decided that it would be wise to leave England. Believing that a public departure would be bad for his business activities, he quietly left for the New World, turning over his business in London to an associate. He traveled to New England and the Caribbean and practiced medicine in South Carolina, where he may have resided for almost three years. After James II's defeat in 1690, Salmon concluded that it was safe for him to return to England. During the early years of the eighteenth century, he began working on his major work, *Botanologia;* he completed it in 1710. In an early section of the herbal, Salmon revealed that he had seen tomatoes growing in Carolina, which was in "the South-East part of Florida." As strange as this may seem today, his geography was accurate because the term *Florida* then referred to what is now the eastern part of the United States. This is the first known reference to the tomato in the British North American colonies.[1]

Several different theories have been espoused to account for the presence of tomatoes in the Carolinas. The most likely explanation is that there were multiple introductions by different peoples at different times for different purposes. The Spanish, who had probably cultivated and consumed tomatoes in their settlements in Florida earlier in the seventeenth century, had established colonies and missions hundreds of miles northward from Saint Augustine, such as Santa Elena on Parris Island. It is probable that the Spanish introduced tomatoes into what is today Georgia and the Carolinas. Alternatively, as gardeners grew tomatoes in Europe, French Huguenot refugees and British colonists may have brought seeds directly to the Carolinas. Finally, tomatoes may also have been introduced from the Caribbean. Migrations from the British West Indies to the southern colonies began in the late seventeenth century and continued throughout the eighteenth century as the rapidly expanding sugar plantations displaced small farmers. Likewise, extensive trade was conducted between the British colonies in the Carolinas and the Caribbean. Part of that trade involved the importation of black slaves from the Caribbean, and these slaves probably had contact with tomato cookery before they were shipped to America. As slaves did the cooking on southern plantations, they may well have introduced and disseminated tomato cookery throughout the South.

While it is certainly not conclusive, linguistic evidence suggests that tomatoes were introduced from the Caribbean. At the time, the term *love apple* was regularly, although not exclusively, used in England; its equivalent was used in France; and the term *tomate* was used throughout Spain's empire. However, almost all eighteenth-century southern references used some variation of the term *tomato.* As previously noted, the word probably originated in Jamaica. By the mid-eighteenth century, it was used throughout the British West Indies.

How local residents in Jamaica pronounced *tomato* in the seventeenth century is unknown. Today, the *a* is clearly pronounced like the *a* in *father* in most English-speaking countries with the exception of the United States. It may have been pronounced this way in British North America as well. In colonial South Carolina the word was spelled *Tomawtoes.* This suggests that the British pronunciation is probably closer to the original than the present-day American one is. Several observers have hypothesized that *tomato* might have come to be pronounced with an *a* as in *bay* because of its similarity to the spelling of *potato.*

THE SOUTHERN STATES

Whatever the initial source, tomatoes were cultivated in the Carolinas by the mid-eighteenth century. Henry Laurens, a prominent South Carolina merchant and plantation owner, grew them at his home in what was then called Charles Town. On July 31, 1764, he sent some to Mrs. Creamer, the wife of the overseer on his plantation. Ten years later the first published references to tomatoes appeared in *The Gardener's Kalendar for South-Carolina* and *The Gardener's Kalendar for North-Carolina.* The North Carolina calendar was based upon the experiences of George Nichols, "whose Garden excelled all others in or near the Latitude of Cape Fear," and Samuel Green, a physician who had developed a physick garden in Wilmington. The author of the South Carolina calendar had "Many Years Experience in the Middle and Southern Parts of the said Province."[2] In both calendars, tomatoes were listed among vegetables and herbs, which indicates that they were not grown just as ornaments.

The author of the South Carolina calendar may have been Martha Logan, a teacher and gardener in Charleston. Her father had emigrated from Barbados. She exchanged seeds, plants, and correspondence with other gardeners and botanists throughout America and Great Britain. She met with Philadelphia naturalist John Bartram, who wrote to the British horticulturist Peter Collinson about her, reporting that Logan's garden was her delight. She published gardening calendars beginning in 1752. Her early calendars did not include tomatoes, but her later ones did. Unfortunately, most of her papers and letters were destroyed.[3]

Only one colonial cookery manuscript is known to have contained a tomato recipe. Its author, Harriott Pinckney Horry, was born in South Carolina in 1748, moved to England, but returned to Charleston in 1763. Five years later she married Daniel Horry. In 1770 she copied a collection of recipes, including one titled "To Keep Tomatoos for Winter Use," which offered instructions on preserving tomatoes so that they could be used for making soup. The recipe was well thought out and reflected a sophistication that indicated it was more than just an experiment. Richard Hooker, a historian and the editor of Harriott Horry's cookery manuscript, speculated that this recipe was derived from her husband's Huguenot family.[4] Huguenots such as Henry Laurens were familiar with tomato culture. However, since tomato cookery was already established in the Caribbean, it was also possible that the recipe derived from Harriott Horry's mother, who had lived in Antigua. As cookbooks often lagged decades behind culinary practices, this recipe shows that tomatoes were used for culinary purposes by some colonial Americans. By the end of the eighteenth century South Carolinians were exporting tomato seeds and cookery to other regions of the United States.

Tomatoes were grown and consumed in other areas of the South. The archaeobotanist Dr. C. Margaret Scarry from the Kentucky Anthropological Research Facility in Lexington reported in 1991 that tomato seeds had been recovered at an excavation of Fort Matanzas near Saint Augustine, Florida. The level at which the seeds were found corresponded to the construction period of the fort (1740–42), indicating that laborers ate tomatoes while working on it. Along Florida's Gulf Coast, tomatoes were used for culinary purposes during the 1780s and 1790s, and by the 1820s they were the most common vegetable grown.[5]

In Georgia, tomatoes were cultivated and consumed in the late eighteenth century and may have been cultivated at an earlier date. By the early nineteenth century Georgians were influencing cookery as far north as Rhode Island. In southern Alabama tomatoes were used to make ketchup during the early nineteenth century and may have been cultivated by French colonists during the eighteenth century.[6]

Despite the tomato's presence along the Atlantic Seaboard and the Gulf Coast, its introduction and adoption in the interior areas of southern states appear to have been delayed. In Salem, North Carolina, tomatoes were not sown until 1833, when a gentleman from South Carolina sent seeds to an "old Mr. Holland." At that time no one had tasted tomatoes, and scarcely any one had heard of them. Similar late arrivals were probably common for other rural areas of the South. In 1820 Phineas Thornton published a thorough inventory of the kitchen garden plants within a twenty-mile radius of Camden, South Carolina, and made no mention of the tomato. His *Southern Gardener and Receipt Book,* published twenty years later, included instructions for cultivating and preparing tomatoes for the table, which suggests that they were introduced in Camden sometime between 1820 and 1840.[7]

Concurrently, as tomato culture expanded in the Carolinas, it also evolved in Virginia. According to Thomas J. Randolph, his grandfather, Thomas Jefferson, asserted that when he was young, tomatoes ornamented flower gardens and were deemed poisonous. By the account of J. Augustin Smith, president of the College of William and Mary, Jefferson met Dr. John de Sequeyra while he was in Williamsburg. Sequeyra had immigrated to America around 1745 and "was of the opinion that a person who should eat a sufficient abundance of these apples, would never die." As he lived to old age, an unusual feat in the marshy environs of Williamsburg, his peculiar constitution supposedly resisted the tomato's deleterious effects. This anecdote was published during Jefferson's lifetime, which offers some credibility on its behalf. Sequeyra's preoccupation with tomatoes was supported by other sources. E. Randolph Braxton reported that "Dr. Secarri," his grandfather's physician, "introduced the custom of eating tomatoes, until then considered more of a flower than a vegetable."[8]

Despite Sequeyra's introduction, it is unlikely that tomatoes were grown extensively in mid-eighteenth-century Virginia. America's first gardening work, John Randolph's *Treatise on Gardening,* written in Williamsburg probably before the American Revolution, made no mention of them. A correspondent in the *Farmer's Register* expounded that tomatoes were hardly ever eaten in Virginia during the 1780s and 1790s. He also explained that Virginians had called them love apples out of ignorance of their proper "foreign title, tomáto," which they pronounced as if it were spelt "chu-mar-tus-iz." While Jefferson was in Paris in the 1780s he sent tomato seeds to Robert Rutherford, who grew and devoured tomatoes in Berkeley County in western Virginia. By 1800 Rutherford had convinced only one other person to eat them.[9]

During the early nineteenth century tomato culture increased. While he was president, Jefferson noted that fresh tomatoes were sold in markets in Washington. They were sold in Alexandria by 1806, which suggests that they were used for culinary purposes by at least some residents. In 1814 they were eaten in Harpers Ferry. In the same year, John James ate them in a public inn near the Natural Bridge in western Virginia. The proprietor claimed that tomatoes had been used as an article of diet in that section as long as she could recollect. Thomas Jefferson grew them at Monticello beginning in 1809 and ten years later served them to Salma Hale. According to Hale, Jefferson claimed to have introduced them into America from Europe. If Hale's recollection was accurate, Jefferson may have been referring to a particular variety of tomato, such as those sent to Robert Rutherford. By the early 1820s they were raised in abundance throughout Virginia and adjoining states and were regarded as a great luxury.[10]

In 1801 Dr. James Tilton found tomatoes growing in Maryland, but they were not used for culinary purposes. Tilton had studied medicine in Europe and was aware of the healthful qualities of tomatoes. He told local residents that they were a vegetable highly esteemed and generally eaten in France, Spain, and Italy. They were especially valuable as a corrective of bile in the system. William Booth sold tomato seeds in Baltimore in 1810. In the following year a Spanish minister found tomatoes growing in Baltimore and remarked that they had been consumed in Spain for many years. By 1817 tomatoes were grown in Chestertown, Maryland.[11]

By 1803 they were sold in the markets in Delaware. Ten years later Mrs. George Read from New Castle compiled a recipe book that contained three tomato recipes. As these recipes were probably copied from an earlier cookery manuscript, tomato cookery was likely to have been in full swing as well in other parts of the Chesapeake region by the first decade of the nineteenth century.[12]

Pennsylvania, New York, and New Jersey

From the southern states, tomatoes spread northward. John Bartram, the preeminent natural scientist in colonial America, found them growing around English-controlled Saint Augustine in 1765. He may have brought their seeds back to sow in America's first botanic garden, which he had established in Philadelphia on the banks of the Schuylkill River. This may not have been Bartram's first contact with tomatoes. He knew and visited Samuel Green in Wilmington and Henry Laurens and Martha Logan in Charleston, all of whom grew tomatoes. He wrote regularly to many of the

European gardeners and botanists, such as Philip Miller, John Hill, Hans Sloane, and Carl von Linne; they had all mentioned tomatoes in their works. On October 10, 1759, Peter Collinson notified Bartram that he did not have a collection of tomatoes but promised to acquire some from James Gordon, a Scottish nurseryman and gardener. This comment was likely to have been in response to a request from Bartram for tomato seeds. Unfortunately, Bartram's original letter to Collinson has been lost. No evidence has been uncovered that Collinson ever sent any tomatoes to Bartram. Tomatoes were listed in the 1807 catalogue for Bartram's garden, but they had been grown and eaten in Philadelphia at least a decade before this date.[13]

Beginning in the late eighteenth century, cookbooks and agricultural books published in Philadelphia contained references to tomatoes, but these were reflective of British experiences, not American. In 1793 Charles Willson Peale, a portrait painter and the creator of one of America's first museums, received "Red Tomato" seeds in a shipment of "a number of subjects of Natural Science" from France. He gave them to his twelve-year-old son, Rubens Peale, who purportedly raised the first tomatoes seen in Philadelphia. About 1795 Rubens's brother, Raphaelle Peale, painted a large globular red tomato in his *Still Life with Vegetables and Fruit.* This is the first American painting to contain tomatoes and one of only four still lifes painted in the United States prior to 1865 known to contain tomatoes. Today it hangs in the Wadsworth Atheneum in Hartford, Connecticut. The other three paintings are in private collections.[14]

Within a few years the tomato was eaten by a segment of the Philadelphia population. In 1804 Dr. James Mease, a physician and botanist, proclaimed that its cultivation was rapidly extending in Pennsylvania, "where a few years ago, it was scarcely known." Eight years later Mease published the first known tomato ketchup recipe. According to the culinary historian William Woys Weaver, Mease's interest in tomatoes may have stemmed from his wife, who was from Georgia. Mease was not the only Philadelphian captivated with tomatoes. John Lithen sold tomato seeds around 1800, as did Bernard M'Mahon. M'Mahon's *American Gardener's Calendar,* published in 1806, stated that the tomato was esteemed for culinary purposes and was "much cultivated for its fruit, in soups and sauces, to which it imparts an agreeable acid flavour." It was also stewed and dressed in various ways and very much admired.[15]

Tomato cultivation rapidly spread throughout Pennsylvania. In 1808 a prisoner grew tomatoes in an enclosure attached to the jail in York, Pennsylvania. When he was discharged, he presented some tomato seeds to the jailor's wife. In 1819 tomatoes were eaten in Buffalo Valley, Pennsylvania, although before this date they had been cultivated there as an ornament. They were also grown ornamentally in Garden Meadows, Lancaster County, before 1820, where children were forbidden to taste them for fear of their having poisonous qualities. While on a visit, Ichabod Allen, a sea captain from New York, saw them growing in the garden. He made a Catalonian salad by "taking the large green bullnose pepper, carefully cutting out the white veins and seeds, and minced with an equal quantity of onion and slicing in the tomato." The salad dressing consisted of oil, egg, mustard, vinegar, and other seasonings. The salad was pronounced very palatable and soon became a favorite dish. In 1822 tomatoes were grown in Octorara, Pennsylvania. In the same year, the *Lancaster Journal* declared that pies made of the "tomatus" were excellent. The editor of the *Lancaster Farmer,* on the other hand, first saw tomatoes under ornamental cultivation in the early 1820s, but since they were considered poisonous, an impression supported by the plant's peculiar odor, he did not venture to taste them until a decade later.[16]

In western Pennsylvania the *Altoona Tribune* reported that tomatoes were disseminated to Frankstown in the spring of 1827 by a traveler who tarried overnight at a village inn. He presented the landlord's daughter with a few seeds, which she planted, and, it was claimed, "from which the first ripe tomatoes ever grown in this country were produced." Here also they were supposed to be poisonous, and the children were cautioned not to touch or handle them. During the following year, however, the girl's mother prepared them for the table, declaring them good and palatable.[17]

A pattern similar to Pennsylvania's occurred in New York. Haitian refugees probably brought tomato seeds to New York during the 1790s. Two New York cookery manuscripts have early recipes for tomato ketchup. The manuscripts were begun in 1795, but the tomato recipes probably were added during the following decade. In 1806 David Hosack listed "Tomatoes or Love Apples" as growing in the Botanic Garden in Manhattan. Within a decade they were consumed in New York City. However, the prominent socialite Charles Haswell stated that, while many New Yorkers grew tomatoes in their gardens, they were universally held to be poisonous and were not essayed as edibles until the early 1820s. It was not until 1826 that Haswell overcame his fear of being poisoned and had the temerity to eat them. Around the same time, Andrew Parmentier, owner of the Brooklyn Horticultural Garden, was convinced that tomatoes were *the* vegetable for America. He offered a dinner featuring tomato dishes, and "the gentlemen and ladies ate them as a compliment to Mr. Parmentier, making as few wry faces as possible."[18]

In upstate New York, tomato cultivation spread northward and westward. By 1819 gardeners in upstate Ogdensburgh grew them, and four years later, gardeners grew them for ornamental purposes in northern St. Lawrence County. In western New York a dinner similar to Parmentier's was served in Rochester in 1825 by a Mr. Tousey, who grew tomatoes from seeds brought from Virginia. Thurlow Weed, a participant in this dinner, was quite incredulous regarding the virtues of the tomato. The guests knew them as love apples but considered them of no value or use. They were not found palatable to any guest at the dinner. This surprised and annoyed their host, who ate them with great gusto. Weed continued to eat them, and when he moved to Albany in 1830, he inquired for tomatoes but was told at the market that they were not produced by the vegetable gardeners and had never been sold in that market. He induced a market gardener to send to New York City for them. The gardener complied, but there were only two customers who bought them. Weed became the editor of the *Albany Evening Journal* and published several timely articles about tomatoes during the mid-1830s.[19]

Despite the presence of tomatoes in Pennsylvania and New York and numerous repetitions of the Robert Gibbon Johnson myth, the earliest primary source pinpointing the tomato in New Jersey was George Perot Macculloch's farm journal, which noted the planting of tomatoes from 1829 onward in Morristown. Tomatoes were also grown in Salem by 1829, and by 1830 they were abundantly grown along the Delaware River valley. James Mapes, from Newark, New Jersey, and the editor of the *Working Farmer,* stated that the tomato was "long grown in our gardens as an ornamental plant, under the name of Love Apple, before being used at all as a culinary vegetable. About 1827 or '28, we occasionally heard of its being eaten in French or Spanish families, but seldom if ever by others."[20]

NEW ENGLAND

In Massachusetts, tomatoes were introduced in the late eighteenth century. In 1797 George Logan, a descendent of Martha Logan, grew tomatoes in Salem from seeds brought from South Carolina. Five years later an Italian painter, Michel Felice Corné, tried to entice Salem residents to eat his beloved *pomidori.* Despite Logan's and Corné's praise, no one was interested. Although cookbooks containing tomato recipes were published in Boston beginning in 1814, these were reprints of British works and did not necessarily reflect conditions in Massachusetts. According to a correspondent in the *Historical Magazine,* sometime between 1815 and 1822 Matthew S. Parker raised the first tomatoes in the southerly part of Boston, in a garden attached to the estate of Warren White. Parker had acquired his seeds from a Mr. Preble, who had a garden outside of Boston. In 1817 they were brought to Plymouth by Dr. Goodwin, who had lived many years in Spain. He planted seeds in the bank garden in Plymouth, and subsequently the plant was distributed throughout the town. The tomatoes were not eaten until 1823 but evidently were not well liked even then. Dr. James Thacher had to reintroduce them in 1831.[21]

In 1819 seeds were sent to Haverhill, Massachusetts, but it was not until two years later that residents found out that tomatoes were good for a salad if cut up and dressed like cucumber. They were exported to Bradford within four years of their arrival in Haverhill. They were served green at the Greenleaf Academy, where they were called *tremarders.* In 1829 B. F. Cutter, a gardener, raised them in Arlington and sold them in a green state for pickles. Later he grew them in Brookline for David Sears, who had learned to use them in France. Tomatoes were not grown in New Bedford until 1830.[22]

While Jared Mansfield served as the U.S. surveyor in Cincinnati early in the nineteenth century, he sent tomato seeds to Wallingford, Connecticut. The resulting fruits were not used for culinary purposes until a visiting woman from New York served them on beefsteaks in 1817. When N. C. Baker from Winchester, Virginia, visited "Bridge Port" in August 1822, he ate the "tomatus." Timothy Dwight, president of Yale, declared that tomatoes grew luxuriantly in 1821 and were cultivated in New England for the table. For Connecticut, these comments may have been premature. Theodore S. Gold, secretary of the Connecticut Board of Agriculture, acknowledged that he grew his first tomatoes in 1832 as a curiosity and made no use of them, although he had heard that the French ate them.[23]

In the early nineteenth century tomatoes were raised ornamentally in Newport, Rhode Island. A gentleman from South Carolina residing in a boardinghouse gave directions for their culinary use, but few tomatoes were eaten in the place for many years except by visitors from Georgia or South Carolina. They were looked upon as curiosities and prized for their beauty. Later they became missiles in the hands of small boys. Another correspondent said that in 1822 he could not find tomato seeds in Bristol, Rhode Island, and had to send off to Philadelphia for some. He moved to Providence, where tomatoes did not become common until 1828.[24]

In 1830, articles in the *Boston Tribune* and the *New England Farmer* reported that the tomato was prized throughout the southern and mid-Atlantic states but said that in New England it had "not won its way to public favor according to its merits." It was said that the tomato was easily raised and could be had from the vine for more than a quarter of a year. The fruit was considered so rich in appearance that it should be cultivated if only for ornament.[25]

Vermont, New Hampshire, and Maine adopted the tomato slightly later than the rest of New England. William Richardson, a member of the U.S. House of Representatives from Massachusetts in 1812, learned to eat tomatoes in Washington, D.C. When he returned home, he brought seeds with him. He sent some to a friend in New Hampshire, where they were raised as curiosities because people were unaware that they were of any use at all for several years. In Londonderry tomatoes were introduced in 1822 by the widow of Rev. William Morrison, who brought seeds from Octorara, Pennsylvania. In Portland, Maine, Joseph Harrod, a dry-goods merchant, grew them from seeds brought from Cuba in 1816. Harrod planted them in the expectation of seeing a nice flowering plant, and it was not until five years later that residents from Haverhill, Massachusetts, told him that they could be eaten. In the spring of 1828, Thomas Fessenden, the editor of the *New England Farmer,* sent tomato seeds to Franklin Glazier in Hollowell, Maine. Glazier grew the first "tomatos" ever seen there. From the fruit, he made ketchup that he preferred to any he had previously purchased. Not everyone in Maine immediately adopted the tomato, however. In 1835 a farmer from New Portland asked the editor of the *Maine Farmer* for a description of the plant, wanting to know where he could find some seeds to cultivate. Twenty years later an article in the same journal claimed that it had been only "a year or two" since the tomato had been first used as food.[26]

THE MISSISSIPPI RIVER SYSTEM AND THE OLD NORTHWEST

As tomatoes spread northward along the Atlantic Coast, they also swept up the Mississippi River system. In Louisiana, Creole, Spanish, Cajun, and French cuisines fused with Native American and African cooking practices to create a culinary caldron that influenced American cookery even before Louisiana was purchased from France by President Thomas Jefferson. The tomato was cultivated and eaten by the French in eighteenth-century New Orleans. Secondary sources avowed that tomatoes were sold in New Orleans markets by 1812. Many early gumbo recipes included them as an ingredient. In the early 1830s, tomatoes were found "bordering on the Mississippi swamp, spreading an unusual length, forming a beautiful vine." The fruit grew in clusters resembling grapes. In 1838 the French-speaking Louisiana gardener J. F. Lelievre affirmed that there were many species of tomatoes used in cooking all of which were equally good.[27]

In the late eighteenth century Colonel Francis Vigo, mentioned in a previous chapter, ate tomatoes in the Old Northwest. Vigo had been born in Sardinia but had joined the Spanish army. He was stationed in Cuba and in 1775 was reassigned to New Orleans. He resigned from the army and became a merchant in Kaskaskia and Vincennes; he helped George Rogers Clark seize the latter town from the British during the American Revolution. Vigo mixed tomato juice with beef gravy and made a kind of ketchup. John Hamtramck, a colonel in the United States Army who commanded the First Regiment, and his wife Rebecca, while garrisoned in Vincennes, adopted the tomato. So did territorial governor and later U.S. president William Harrison, who resided in Vincennes. When the Hamtramcks were reassigned in 1798, they took tomato seeds with them to Detroit and later to Fort Wayne, Indiana. Colonel Hamtramck died in 1803, and his wife remarried Jesse Thomas, who later became a U.S. senator. She served tomatoes to him in Lawrenceburg, Indiana, where they lived in 1807.[28]

Tomatoes were grown in Ohio by the turn of the nineteenth century. Thomas Ewing, later a U.S. senator from Ohio, remembered that in the summer of 1800, Apphia Brown ate a love apple. Ewing, then eleven, and his playmates passed this fearful intelligence to the grownups, who did not partake of their alarm, and it passed off without a catastrophe. Seth Adams, a prominent Ohio farmer, recounted that his brother, who had been introduced to tomatoes in Louisiana, came up the Mississippi River to Ohio and gave him some seeds in 1808. After they grew into vines and produced fruit, Adams boiled the tomatoes, squeezed out the juice, and put them on bread, but the dish so prepared was not palatable. Other Ohioans ate them raw and knew how to cook them. They were sold in Cincinnati markets by 1813. When Governor De Witt Clinton of New York visited Chillicothe in July 1825, the Madeira Hotel offered *tomattoes* as a vegetable. When the British author Frances Trollope visited Cincinnati in 1832, tomatoes were available in markets from July to December. She reported that they were the great luxury of the American table, in the opinion of most Europeans.[29]

In Lexington, Kentucky, by 1817 children freely ate tomatoes right from the vine in the garden. Two years later, Henrico M'Murtrie recorded that tomatoes were growing in Louisville. In Illinois, French settlers grew tomatoes in Kaskaskia by 1807 or 1808. They were likely brought up the Mississippi River from New Orleans. In southern Illinois tomatoes were grown by 1829. In 1833 German immigrants found them growing on a farm in St. Clair, Illinois, "though the value of this delicious fruit was then unknown to us and therefore not appreciated; in fact, tomatoes were considered by the new comers as unwholesome and even poisonous."[30]

Colonel Hamtramck's introduced tomatoes to Detroit before the close of the eighteenth century. By 1832, the Ladies of the Detroit Free School Society prepared tomato ketchup bottles by the dozen and sold them at E. Bingham's store to raise funds for the school. Tomatoes were first grown in Grand Rapids, Michigan, decades later. Abel Page grew them in 1836, but only a schoolteacher would eat them. During the following year, the *Detroit Spectator* published an article on the healthful virtues of the tomato, but the editor reported years later that, though many persons tried the vegetable, "it was a long time afterward, ere the tomato could be eaten readily. Many scoffed at the idea of its ever being made an article of food." Despite this gloomy perspective, the *Michigan Farmer* praised the tomato with articles about its cultivation and numerous recipes from its inception in 1842.[31]

As the rest of the West was settled, tomatoes moved along with the farmers. Tomato products were sold in Davenport, Iowa, by 1838. The *Iowa Farmer* included articles about tomatoes from its inception in 1853. Tomatoes were eaten in Madison, Wisconsin, by 1840. The *Wisconsin Farmer* included articles about tomatoes from its establishment in 1849. By the mid-1850s, however, English and Scotch settlers in Wisconsin still viewed the tomato with contempt. In Minnesota, tomatoes were cultivated as early as 1846. Further south in what was then Indian Territory, tomato ketchup was probably consumed at Fort Scott in Kansas by 1842, and by 1845 it was sold at Fort Gibson in what is today Oklahoma.[32]

CALIFORNIA, THE AMERICAN SOUTHWEST, AND THE PACIFIC

Tomatoes were grown in California and in areas that make up the American Southwest well before the acquisition of these areas by the United States. Franciscan priests, who established most of the missions in the Spanish-controlled parts of America, grew tomatoes in their gardens. By 1763 they were used for medicinal purposes in Sonora, part of which is today in the state of Arizona. In California Fr. Fermín Francisco de Lasuén, stationed at Mission San Diego, received a donation of tomatoes from Mission San Gabriel near Los Angeles in 1777. In 1810 Fr. Antonio Cavallero from the Cochiti Pueblo in what is today the state of New Mexico recorded that his tomatoes had been destroyed by hail. They were grown in Texas by the early nineteenth century if not before.[33] As few such references have been located in Spanish America, the tomato was probably of minor culinary importance there.

Tomatoes were also brought to the Pacific Northwest and Hawaii by missionaries and settlers from the eastern United States. Oregon became a territory of the United States after the Louisiana Purchase in 1803. By 1839 tomatoes were growing on the banks of the Wallawalla River. Two years later Lieutenant Johnson found them growing at the Kooskoosky Mission Station and speculated that they had been imported from the eastern United States. The historian Sidney Warren maintained that pioneers in Oregon and Washington boiled tomatoes with sugar until they became very thick, then added honey and placed the concoction upon hotcakes. He also claimed that the pioneers boiled them, dried them in the sun, coated them with sugar, and ate them as a confection. Although Hawaii did not become an American territory until 1898, missionaries from New England who were there in the 1830s wrote home about tomatoes that grew year-round in their gardens.[34]

THE BRITISH, FRENCH, AND CREOLE INFLUENCES

Though the tomato's introduction process was haphazard, two major influences were particularly important in the early expansion of tomato culture and cookery in America. The first was British. From the Old World, colonists brought agricultural and horticultural practices, cookery books, and medical theories. Long after the American Revolution, British cultural influences continued to dominate the newly independent nation. British gardening books, shipped from England or reprinted in America, incorporated references to tomatoes. Charles Varlo, a farmer and agricultural writer in Scotland and Ireland, ventured to America in 1785 and published in Philadelphia an edition of his *New System of Husbandry* that mentioned tomatoes. In 1799 an American edition of Charles Marshall's *Introduction to Gardening* was published in Boston. This gardening book mentioned red, yellow, white, and cherry-fruited tomato varieties and reported that red and green varieties were good for pickles while the red variety strengthened soups. In 1803 an American reprint of the British work *Gleanings from the Most Celebrated Books on Husbandry, Gardening and Rural Affairs* noted two varieties of tomatoes, both of which were prescribed for medicine and also used in sauces, soups, and pickling.[35] Despite the publication of these works in America, they reflected British and Scottish usage of the tomato, not American.

English practices also dominated the American colonial kitchen. Colonists brought cookery books and manuscripts with them to the New World. In addition, cookbooks were shipped regularly from England, and many were reprinted in America. Even after the American Revolution, British cookbooks controlled the American market. Richard Briggs, a British tavern owner, revised Hannah Glasse's previously cited recipe for haddock with love apples and incorporated it into his cookbook, which was published in Philadelphia in 1792 and again in 1798. As culinary historian William

Woys Weaver has mentioned, Briggs's cookbook was popular among Philadelphia Quakers, and through Mrs. Goodfellow's cooking school, its influence went far beyond those who owned the book.[36] What affect, if any, this recipe had upon tomato cookery in America is unknown.

The British influence was not limited to published works. British and Irish immigrants brought knowledge of tomato culture and usage to America. London-born Dr. John de Sequeyra, whose family was of Portuguese-Jewish origin, immigrated to America around 1745 and introduced the tomato to Williamsburg, Virginia, as noted earlier, and Liverpool-born Dr. Samuel Green planted tomatoes in his physick garden in Wilmington prior to 1771. Robert Squibb migrated from England to Charleston in 1780 and, seven years later, published extensive directions for growing tomatoes in his *Gardener's Calendar for North-Carolina, South-Carolina and Georgia*. In Washington, D.C., David Hepburn, a gardener who had twenty years of gardening experience in Britain before emigrating to America around 1784, grew tomatoes on Mason's Island in the Potomac. Hepburn, along with John Gardiner, published *The American Gardener* in 1804; it was the first indigenous book published in the United States devoted solely to horticulture. It offered instructions for the cultivation of tomatoes.[37]

In Philadelphia, English immigrants Cuthbert and David Landreth received tomato seeds from Rubens Peale during the 1790s and may have sold the fruit to French immigrants, but there was evidently little demand from others. Irish immigrant Bernard M'Mahon sold seeds for "tomatoes or love apples" in Philadelphia in 1800. Scottish-born Grant Thorburn sold "love apple" seeds in New York in 1807.[38] The Landreths, M'Mahon, and Thorburn not only sold tomato seeds but also promoted tomato culture and cookery. Many seedsmen incorporated directions for tomato cultivation and cookery in their seed catalogues or gardeners' calendars. Seedsmen and gardeners communicated with botanists and other natural scientists in America and Europe. Knowledge about the tomato flowed through the seedsmen and gardeners to local farmers.

The second major influence on tomato cultivation and cookery in America was of French and Creole derivation. This germinated in several different ways. Huguenot refugees, other French settlers, and their progeny in South Carolina, New Orleans, and the Old Northwest cultivated and consumed tomatoes beginning in the eighteenth century. As previously remarked, Jefferson sent tomato seeds from France to Virginia during the 1780s. Tomato seeds from France were exported into Philadelphia in 1793. French refugees who flocked to the United States during the excesses of the French Revolution have been credited with introducing the tomato to

America. Whether or not French restaurants served tomatoes in the eighteenth century, they offered them prodigiously during the nineteenth century in America.

In addition to the direct influence from metropolitan France, American tomato cookery was influenced by French and Creole refugees from what is today Haiti. Inspired by the French Revolution and led by Toussaint L'Ouverture, Haitian slaves revolted in 1791. Many French refugees, along with their slaves and servants, immigrated to America during the following decade. When Peale grew tomatoes in Philadelphia in the 1790s, he grew them in hopes of raising a flower. A French gentleman from Haiti recognized the "Tomato as a favorite fruit of his." According to Weaver, this may have been a reference to Pierre Bossée, an innkeeper from Cap Française in Haiti, who opened a boardinghouse and restaurant in Philadelphia shortly after his arrival in 1793. Three years later a Haitian refugee named Nicalo arrived in Philadelphia, bringing seeds from several vegetables, including the tomato. When his tomatoes ripened, Nicalo dressed them in a salad, which was relished by the neighboring Eldridge family. Other neighbors raised them as ornaments, having the impression that they were poisonous.[39]

French and Creole refugees from Haiti also cultivated tomatoes in Maryland and other places near the shores of the Chesapeake Bay. John James believed that Haitian refugees brought tomatoes to Alexandria. These refugees were of course well acquainted with the tomato. James always supposed that they had introduced it into the United States. Outen Laws, from Wilmington, Delaware, remembered that in 1803 only a few bushes of the tomato were grown in his flower garden, and their fruits were conspicuously arranged on the table, their "bright hue attracting the admiration of the visitor." Refugees from Haiti settled in Delaware, together with their black serving women, companions of their flight. These "colored women" possessed refined manners and were regarded as artistic cooks. Laws believed that these women, "by the exercise of their arts, in the preparation of tomato for the table, may have overcome our aversion to it as an article of food."[40]

Americans of French descent not only ate tomatoes but also introduced them to other countries. John James Audubon was born in New Orleans; his father was an officer in the French navy and his mother a Creole. They moved to Haiti, and his mother died there during the slave uprising. Audubon moved to France, studied ornithology, and returned in 1803 to America, where he began painting birds. In 1826, while visiting a Mr. Roscoe in Liverpool, England, Audubon ate a raw tomato and quite astonished his hosts.[41]

While British influence proportionally declined as decades of the nineteenth century passed, many British recipes and gardening articles continued to be published in America throughout the mid-nineteenth century. As British influence waned, French influence on tomato cookery waxed. French cookery books began to be published in America in 1828. Throughout the 1830s, 1840s, and 1850s French restaurants in most large cities extolled the virtues of the tomato.

WHY MANY AMERICANS DID NOT EAT TOMATOES

Despite widespread diffusion of the tomato in the early nineteenth century, there were several reasons why many Americans did not eat them in addition to the belief that they were poisonous. Many colonists of English, Scottish, and Irish heritage had emigrated to America before the tomato became a common culinary product in Britain and Ireland. The physical isolation of America during the late eighteenth and early nineteenth centuries contributed to the lack of familiarity with the growing popularity of the tomato in Europe. As late as 1835, farmers requested descriptions of the tomato plant, sought instruction as to where seeds could be obtained, and asked for information on the tomato's cultivation and usage. Others knew how to cultivate tomatoes but possessed no knowledge about how to cook them because few tomato recipes had been published prior to the 1820s and few cookery manuscripts contained references to them.

Besides the fear specifically of tomatoes, there remained the ancient fear of vegetables in general. The historian Richard Hooker suggested that this may have persisted because of the practice of washing vegetables in polluted water or because they were eaten during the summer, when epidemics were common. And while some Americans obviously did believe that tomatoes were poisonous, this phenomenon has been blown out of proportion by well-intentioned popular historians. Research has located only three references to the tomato's purported poisonous qualities published in America prior to 1860. One was from a reprinted British medical work, reflecting outmoded beliefs in Britain rather than in America. The second was a facetious comment published in a newspaper. The third was a comment by Thomas Jefferson's grandson, indicating that the tomato was considered poisonous when Jefferson was young. About twenty-five plausible references to its purported poisonous qualities were published after 1860 in autobiographies or reminiscences. These retrospective accounts attempted to explain the discrepancy between the tomato's lack of previous usage juxtaposed to its then current general consumption. The drama associated with the preposterous notion of the poisonous tomato may have been the major factor in the repetition of these tales.[42]

There were other, more significant, if mundane, reasons why Americans chose not to eat tomatoes. Some did not like the smell of the tomato. To a farmer's wife, the plant smelled nauseous. To Philip R. Freas, editor of Philadelphia's *Germantown Telegraph,* tomatoes were "a very poor garden decoction, *with a bad smell.*" A woman in upstate New York cultivated them "as an ornamental plant, and would as soon think of cooking ripe potato balls," adding that the odor of the plant was of itself a sufficient warning against its fruit. A correspondent to the *Genesee Farmer* urged readers to make piquant tomato sauce so it would retain less of the essential odor of the plant. To counteract this unpleasant smell, some tomato recipes were aromatized with herbs and spices. Concern with the smell of the plant continued throughout the nineteenth century, even after the tomato became one of the most important vegetables in America.[43]

Others did not like the appearance or the taste of the tomato. A gardener in Massachusetts stated that the first time he saw tomatoes was during the 1820s. They "appeared so disgusting that I thought I must be very hungry before I should be induced to taste them." The look of the tomato was so disagreeable that many people supposed that it would "never receive a permanent place in our list of culinary vegetables." In Pennsylvania during the late 1820s, J. B. Garber posited that "hardly two persons in a hundred, on first tasting it, thought that they would ever be induced to taste that *sour trash* a second time."[44]

This dislike of the tomato was not limited to northerners, though the fruit was eaten more widely in the South. Charles G. Brietz from North Carolina, while traveling north in 1831, described a dinner at a public inn outside Richmond. He was served some sliced fruit with a beautiful red color, "thereby spoiling my entire dinner, and being too green and bashful to have my plate removed and take another, I did without much dinner. This was my first introduction to tomatoes, and I must say, that I do not like them much to this day." Brietz wrote that statement in 1887—fifty-six years after his first experience with them. A similar experience befell S. D. Wilcox, the editor of the *Florida Agriculturist,* who reported that in 1836 he ate his first tomatoes as an experiment in the form of a pie without spices or sugar. After consuming it, he believed that the tomato "was an arrant humbug and deserved forthwith to be consigned to the tomb of all the Capulets." At that time, according to Wilcox, anyone "who would have predicted that the tomato would ever become popular as an esculent or to be used in any utilitarian way except as a gratification to the eye, would have been set down at once as daft or visionary."[45]

A correspondent in the *Genesee Farmer* surmised that initially people ate tomatoes for their healthful effects, but most became fond of them after eating them a few times. The editor of the *American Farmer* theorized that to most people the flavor was disagreeable at first, but a little use entirely counteracted or removed that effect. In 1834 the *Cincinnati Farmer and Gardener* too maintained that the acid flavor became agreeable to most persons, although it was not always relished when first tasted. In 1835 Charles Crosman, a gardener and seedsman, wrote that there were "but few who relish the tomato at the first taste." In 1847 cookbook author Eliza Leslie believed tomatoes had a raw taste that was unpleasant to most people. In the mid-1850s W. W. Hall, editor of the *Journal of Health,* surmised that there were still many people who thought the flavor of the tomato unpleasant, and still others thought tomatoes disgusting. Scottish-born John Muir, soon to become one of America's preeminent naturalists, reported that in Wisconsin in the mid-1850s, English and Scottish settlers had nothing but contempt for tomatoes, which were "so fine to look at with their sunny colors and so disappointing in taste." As late as 1856 Ralph Waldo Emerson concurred with many other observers that the taste for tomatoes was an acquired one.[46]

Other comments were less polite. A writer in the *Genesee Farmer* stated that the tomato was offensive in whatever shape it was offered. A writer in the *Horticulturist* stated that tomatoes were "odious and repulsive smelling berries." Even these comments paled by comparison to those of Joseph T. Buckingham, editor of the *Boston Courier,* who stated that they were "the mere fungus of an offensive plant, which one cannot touch without an immediate application of soap and water with an infusion of *eau de cologne,* to sweeten the hand—tomatoes, the twin-brothers to soured and putrescent potato-balls—deliver us, O ye caterers of luxuries, ye gods and goddesses of the science of cookery! deliver us from tomatoes."[47] After this statement Buckingham bowed to the dictates of fashion and meekly offered his readers two tomato recipes.

Still others were concerned about the nutritional value of the tomato. In 1838 William Alcott claimed that no one believed the tomato was nutritious and that some persons were even injured by its acid. A correspondent in the *Maine Farmer,* identified only as "Candor," concurred with Alcott. According to Candor, medicine and food were two different things, and they should not be mixed. In proportion to the fitness of an article to be used as a medicine was its unfitness to be used as food. There was simply no call for it as the range of selection of other fruits and vegetables was already so wide. Why meddle with the doubtful tomato, which "needs so much effort at improvement?" The editors of the *Yankee Farmer* responded by stating that Candor's comments were, in their humble opinion, one of the most absurd statements that was ever made. They believed that they were "supported by correct observation, experience, and common sense, and none who honor the profession of medicine will . . . advance a different opinion."[48]

One physician who did advance a different opinion was Dio Lewis. Lewis had been trained as a physician at Harvard's medical school and was one of the founders of the physical culture movement in the United States. While he was a youth, his mother advised him not to handle tomatoes, and he remembered the injunction well. In 1852 he began a food lecture tour. After a presentation in Cincinnati in which he criticized tomatoes, seven people came up on the platform and testified to having had a peculiar condition of the stomach, developed when the tomato season first began. Lewis claimed to know a number of cases of piles caused by excessive use of tomatoes and to know "many persons to suffer from tender and bleeding gums, from 'teeth set on edge,' and quite a number from loose teeth, produced by eating tomatoes." He related the sad story of a young lady who visited her uncle in the country. She was told that tomatoes were the healthiest vegetables she could eat. She soon learned to like them and picked them right off the vines. She said that

> Almost immediately my mouth became sore, and my gums bled freely upon the use of the tooth-brush. But I was told this was a disease in my stomach working off through my mouth. No one suspected the tomatoes. When I came home, I brought with me a bushel and a half, and ate them as long as I could preserve them. In the mean time my teeth had become loose. At length they became so very loose that I began to take them out with my fingers, and I now have but one tooth left, and if you would like to have me take that out, I can do it with my fingers.

Lewis's conclusion was that if people were fond of tomatoes they could "eat them in small quantities, say one or two teaspoonfuls of cooked tomatoes at a meal, as a sauce." He believed that the tomato was "*medicinal,* and should never be used in any considerable quantity by healthy people." Tomatoes should be placed "in the category with medicines, and prescribed when necessary by a medical man."[49]

PATTERNS OF INTRODUCTION AND ADOPTION

There were several different processes of adoption in America during the early nineteenth century. The tomato was initially a product of cities, with the exception of the areas controlled by the Spanish, and only later did it penetrate into the rural hinterland. This is not surprising. People who lived in urban areas were more exposed to outside influences. In the early nineteenth century most cities were located along the Atlantic Ocean, the Gulf Coast, or navigable rivers such as the Mississippi. Refugees and settlers, some of whom had already adopted the tomato, entered America through cities, and many remained in or near them. Likewise, those engaged in trade with regions of the world that had already adopted the tomato, such as the islands of the Caribbean, operated through these cities. Conversely, those who settled in rural areas were more conservative in their willingness to try new foods. As the tomato was so unlike any other commonly used fruit or vegetable in America, it was even less likely to be readily adopted. There were ample other available fruits and vegetables, as observers noted, and thus there was no particular need for the tomato.

The tomato's introduction into a specific community did not necessarily mean its immediate adoption as a culinary ingredient. Tomatoes were grown in northern gardens as ornaments or curiosities years before they were timidly tasted. Likewise, their culinary usage by some segments of a community did not mean that everyone used them for culinary purposes. French and Creole refugees ate tomatoes in Philadelphia long before other Philadelphians deigned to do so. Many northerners were aware that the French, Italians, and Spanish ate them, but this knowledge was not sufficient to encourage them to emulate this behavior.

These social and cultural components of diffusion and adoption were mediated by a variety of other factors, such as regional climatic conditions, local soil composition, knowledge of cultivation techniques, the availability of seeds, and the understanding of culinary techniques. In parts of the southern states, tomatoes grew almost spontaneously. In northern and midwestern states, they could be easily grown only if the farmer was aware of how to cultivate them. Also, the growing season was shorter in the North than in the South. These factors contributed to the tomato's early success in the southern states and its later adoption in the North.

NOTES

1. William Salmon, *Botanologia. The English Herbal or, History of Plants* (London: Printed by I. Dawks for H. Rhodes and J. Taylor, 1710), 29; St. Julien Childs, "The Ubiquitous Dr. Salmon," *The Journal of the South*

Carolina Medical Association, 66 (May 1970): 160–61; C. J. S. Thompson, *Quacks of Old London* (New York: Barnes & Noble, 1993), 128–29.

2. George C. Rogers, Jr., ed., *The Papers of Henry Laurens* (Columbia: University of South Carolina Press, 1974), vol. 4, 359; George Andrews, *Well's Register: Together with an Almanack* (Charlestown: Robert Wells, 1774), 15, 17.

3. Martha Logan, *The South-Carolina Almanack, for the Year of our Lord 1756; and the Palladium of Knowledge: or the Carolina and Georgia Almanac, for the Year of our Lord 1798* (Charleston, S.C.: National Society of the Colonial Dames of America in the State of South Carolina, 1976).

4. Harriott Pinckney Horry Papers, Receipt Book, 28, South Carolina Historical Society #39–19; Richard Hooker, ed., *A Colonial Plantation Cookbook: The Receipt Book of Harriott Pinckney Horry, 1770* (Columbia: University of South Carolina Press, 1984), 25, 89.

5. C. Margaret Scarry, "Plant Remains from Fort Matanzas," unpublished paper dated February 1991; Marion Francis Shambaugh, "The Development of Agriculture in Florida During the Second Spanish Period," master's thesis, University of Florida, 1953, 71; John Lee Williams, *A View of West Florida* (Philadelphia: H. S. Tanner and the Author, 1827), 67; John Williams, *The Territory of Florida: Sketches of the Topography, Civil and Natural History* (New York: A. T. Goodrich, 1837), 24.

6. *New England Farmer,* 14 (October 14, 1835): 106; *Country Gentleman,* 19 (April 10, 1862): 239; Mrs. Gaines diary, as in the *Mobile Daily Register,* July 6, 1879.

7. Eliza Vierling Kremer, "Bits of Old Salem Gossip," manuscript at the Salem College Library, Winston-Salem, N.C.; *American Farmer,* 2 (November 1820): 263; P. Thornton, *The Southern Gardener and Receipt Book* (Camden, S.C.: Printed for the Author, 1840), 55.

8. J. Austin Smith, as in Thomas Sewall, *A Lecture Delivered at the Opening of the Medical Department of the Columbian College, in the District of Columbia* (Washington: Printed at the Columbian Office, March 30, 1825), 61; Thomas J. Randolph, "An Essay Delivered before the Agricultural Society of Albemarle at their Annual Fair, on the 29th, of October, 1842" (Charlottesville, Va.: James Alexander), 7; Accession File #56.568 for the *Portrait of Dr. John de Sequeyra* at the Winterthur Museum in Delaware; Robert Shosteck, "Notes on an Early Virginia Physician," *American Jewish Archives,* 23 (November 1971): 198–212.

9. Marjorie Fleming Warner, "The Earliest American Book on Kitchen Gardening," *Annual Report of the American Historical Association for the Year 1919* (Washington: Government Printing Office, 1923), vol. 1, 433–42; *Farmer's Register,* 9 (September 1841): 589; Thomas Brown, "Account of the

Lineage of the Brown Family," manuscript dated 1865, 48–49, Ambler-Brown Family Papers at the Special Collections Library, Duke University Press, Durham, N.C.

10. Henry S. Randall, *The Life of Thomas Jefferson* (New York: Derby & Jackson, 1858), vol. 1, opposite p. 44; Thomas Jefferson, *Thomas Jefferson's Garden Book 1766–1824 with Relevant Extracts from his other Writings,* annotated by Edwin Morris Betts (Philadelphia: American Philosophical Society, 1981), 391, 403, 639; *Historical Magazine,* 6 (March 1862): 102; *Reporter and Watch Tower,* August 19, 1885; *Country Gentleman,* 38 (June 5, 1873): 366; 38 (December 25, 1873): 820; *Vincennes Western Sun,* April 17, 1874; *Proceedings of the Massachusetts Historical Society,* 46 (1913): 408; Sewall, *Lecture,* 61–62.

11. *Reporter and Watch Tower,* August 19, 1885; *Country Gentleman,* 38 (June 5, 1873): 366; William Booth, *A Catalogue of Kitchen Garden Seeds and Plants* (Baltimore: G. Dobbin and Murphy, 1810), 4; W. Emerson Wilson, ed., *Plantation Life at Rose Hill: The Diaries of Martha Ogle Forman 1814–1845* (Wilmington, Del.: Historical Society of Delaware, 1976), 42.

12. Manuscript cookbook, Mrs. George Read, New Castle, 1813, Holcomb Collection, Historical Society of Delaware, Wilmington.

13. Edmund Berkeley and Dorothy Smith Berkeley, *The Correspondence of John Bartram 1734–1777* (Rueful: University Press of Florida, 1992), 473; John Bartram, "Diary of the Journey through the Carolinas, Georgia and Florida from July 1765 to April 10, 1766," annotated by Francis Harper, *Transactions of the American Philosophical Society,* n.s., part 1, 33 (1942), 53; John Bartram, *A Catalogue of Trees, Shrubs, and Herbaceous Plants* (Philadelphia: Bartram and Reynolds, 1807), 33.

14. Rubens Peale, *Memorandum and the Events of His Life,* vol. 1, back of p. 4, manuscript in the Peale-Sellers Papers at the American Philosophical Society, Philadelphia; *Still Life with Vegetables and Fruits,* painted by Raphaelle Peale, at the Wadsworth Atheneum in Hartford, Conn.; letter from Elizabeth Kornhauser, Curator, Wadsworth Atheneum, to the author, dated July 13, 1993.

15. A. F. M. Willich [James Mease, ed.], *The Domestic Encyclopedia* (Philadelphia: William Young Birch and Abraham Small, 1804), vol. 3, 506; Mease, ed., *Archives of Useful Knowledge* (Philadelphia: David Hogan, 1812), vol. 2, 306; John Lithen, *Catalogue of Garden Seeds* (Philadelphia: 1800); Bernard M'Mahon, *A Catalogue of Garden, Grass, Herb, Flower, Roots* (Philadelphia: 1800); M'Mahon, *The American Gardener's Calendar; Adapted to the Climates and Seasons of the United States* (Philadelphia: Printed by B. Graves for the Author, 1806), 319.

16. *Lancaster Farmer,* 14 (November 1882): 161; John Blair Linn, *Annals of Buffalo Valley, Pennsylvania, 1755–1855* (Harrisburg, Pa.: Lane S. Hart, 1877), 443; *Willey's Book of Nutfield: A The History of that Part of New Hampshire Comprised within the Limits of the Old Township of Londonderry* (Derry Depot, N.H.: George F. Willey, 1895), 102; *Lancaster Journal,* September 6, 1822.

17. *Altoona Tribune,* as in the *Lancaster Farmer,* 14 (November 1882): 161.

18. "Approved Recipes," 1795, 77, and "Receipt Book of Sally Bella Dunlop; also James Dunlop," 1795, cookery manuscripts at the New York State Historical Association; David Hosack, *A Catalogue of Plants Contained in the Botanic Garden at Elgin* (New York: 1806), 26; Richard Alsop, *The Universal Receipt Book or Complete Family Direction by a Society of Gentlemen in New York* (New York: I. Riley, 1814), 45; *New York Observer,* 45 (February 28, 1867): 72; Charles Haswell, *Reminiscences of an Octogenarian of the City of New York* (New York: Harper & Brothers, 1896), 148.

19. *Country Gentleman,* 39 (October 1, 1874): 630; 19 (May 15, 1862): 318; Harriet Weed, ed., *Autobiography of Thurlow Weed* (Boston: Houghton, Mifflin, 1883), 205; *Albany Evening Journal,* August 14, 1835; September 17, 1835.

20. *Country Gentleman,* 19 (May 15, 1862): 318; George Perot Macculloch's Ledger, manuscript at Macculloch Hall Historical Museum, Morristown, N.J.; *Boston Tribune,* as in the *New England Farmer,* 9 (August 6, 1830): 20; *New England Farmer,* 9 (August 27, 1830): 45; Charles W. Casper, manuscript report on canning in Salem County, Salem County Historical Society, dated March 13, 1906; *Working Farmer,* 4 (January 1, 1853): 244.

21. William Bentley, *The Diary of Reverend William Bentley, D. D., Pastor of the East Church of Salem, Massachusetts* (Glouster, Mass.: Peter Smith, 1982), vol. 2, 240–41, 453; *Historical Magazine,* 6 (January 1862): 35–36; *Reporter and Watch Tower,* August 19, 1885; William Davis, *Plymouth Memories of an Octogenarian* (Plymouth, Mass.: Memorial Press, 1906), 508.

22. *Country Gentleman,* 38 (December 11, 1873): 790; 19 (April 10, 1862): 239; *New England Farmer,* 2d ser., 1 (December 1867): 561.

23. *Western Reserve Magazine of Agriculture and Horticulture,* 1 (July 1845): 100; Timothy Dwight [Barbara Miller Solomon and Patricia M. King, eds.], *Travels in New England and New York* (Cambridge, Mass.: Harvard University Press, 1969), vol. 1, 29–30; N. C. Baker, Memorandum Book, 1822, Special Collections Library, Duke University, Durham, N.C.; *Country Gentleman,* 45 (May 13, 1880): 310–11.

24. *Boston Evening Transcript,* January 20, 1879; *Country Gentleman,* 19 (April 10, 1862): 239.

25. *Boston Tribune,* as in the *New England Farmer,* 9 (August 6, 1830): 20; *New England Farmer,* 9 (August 27, 1830): 45.

26. Letter dated November 29, 1828, as in the *New England Farmer,* 7 (December 12, 1828): 167; 2d ser., 1 (December 1867): 561; *Willey's Book of Nutfield: A The History of that Part of New Hampshire Comprised within the Limits of the Old Township of Londonderry* (Derry Depot, N.H.: George F, Willey, 1895), 102; *Portland Daily Advertiser,* September 10, 1873; *Maine Farmer,* 3 (October 16, 1835): 289; 21 (September 1, 1852).

27. *Letter Diary of Joseph Delfau de Pontalba to Wife,* May 8, 1796, September 17, 1796, Louisiana State Museum Historical Center, New Orleans; *Boston Evening Transcript,* March 31, 1879; Basil Hall, *Travels in North America, including a New Preface by Ferdinand Anders and 12 Contemporaneous Illustrations* (Edinburgh: Cadill, 1829), vol. 3, 332; *Southern Agriculturist,* 5 (December 1832): 663; J. F. Lelievre, *Nouveau jardinier de la Louisiane,* as translated by George McCue, "The History of the Use of the Tomato: An Annotated Bibliography," *Annals of the Missouri Botanical Garden,* 39 (November 1952): 343.

28. *Western Reserve Magazine of Agriculture and Horticulture,* 1 (July 1845): 100; *Vincennes Weekly Western Sun,* June 18, 1864; *Western Sun,* April 17, 1874; David Thomas, *Early Travels in the Western Country in the Summer of 1816* (Auburn, N.Y.: David Rumsey, 1819), 165.

29. Letter from Thomas Ewing to the Athens County Pioneer Association, dated July 3, 1871, as in "The Autobiography of Thomas Ewing," edited by Clement C. Martzolff, *Ohio Archæological and Historical Quarterly,* 22 (January 1913): 187; *Western Reserve Magazine of Agriculture and Horticulture,* 1 (July 1845): 100; Charles Sumner Plumb, "Seth Adams A Pioneer Ohio Shepherd," *Ohio Archæological and Historical Quarterly,* 43 (January 1934): 14–15; *Country Gentleman,* 38 (December 25, 1873): 820; Madeira Hotel, Bill of Fare of a Dinner Given by the Citizens of Chillicothe, and its Vicinity in Honor of the Guest of Ohio, Gov. De Witt Clinton of N. York, July 25, 1825, New-York Historical Society; Frances Trollope, *The Domestic Manners of the Americans. Edited, with a History of Mrs. Trollope's Adventure in America by Donald Smalley* (New York: Knopf, 1949), 61.

30. *Boston Evening Transcript,* February 3, 1879; Henrico M'Murtrie, *Sketches of Louisville* (Louisville, Ky.: S. Penn, 1819), 226; *Western Reserve Magazine of Agriculture and Horticulture,* 1 (July 1845): 100; Milo Milton Quaife, ed., *Growing Up in Southern Illinois 1820 to 1861; Memoirs of Daniel Harmon Brush* (Chicago: Lakeside Press, R. R. Donnelley, 1944), 53; Thomas Joseph McCormack, ed., *Memoirs of Gustave Kroerner* (Cedar Rapids, Iowa: Torch Press, 1902), vol. 1, 300.

31. Silas Farmer, *History of Detroit and Wayne County and Early Michigan,* 3d ed. (Detroit: Silas Farmer, 1890), 736. Allen Baxter, *History of the City of Grand Rapids* (New York & Grand Rapids: Munsell, 1891), 506; *Old Colony Memorial,* September 8, 1849; *Life in the West* (London: Sounders and Otley, 1842), 266–67; *Michigan Farmer,* 1 (May 13, 1843): 54; (June 13, 1843): 67; (July 1, 1843): 75; (July 15, 1843): 87.

32. *Iowa Sun and Davenport & Rock Island News,* August 11–November 17, 1838; Rodney Loehr, *Minnesota Farmer's Diaries William R. Brown, 1845–46; Mitchell Y. Jackson, 1852–63* (Saint Paul: Minnesota Historical Society, 1939), 75, 80; John Muir, *The Story of My Boyhood and Youth* (Boston & New York: Houghton Mifflin, 1912), 210; *Iowa Farmer and Horticulturist,* 1 (June 1853): 21, 63, 122, 133; "Council of Administration Minutes for Ft. Gibson," May 8, 1845, as in David Delo, *Post Traders and Post Traders: The Army Sulter and the Frontier* (Salt Lake City: University of Utah Press, 1992), 52.

33. "The Rudo Ensayo," *Records of the American Catholic Historical Society of Philadelphia,* 5 (June 1894): 156; letter of Fr. Antonio Cavallero, Cochiti Pueblo, dated August 8, 1810, as translated and published by *The Spanish Archives of New Mexico,* 2d ser. (Santa Fe), document #2350; "Thomas Eastland Diary of 1849," *California Historical Society Quarterly,* 23 (June 1939): 127; John Fox Hammond, *A Surgeon's Report on Socorro, New Mexico, 1852* (Santa Fe: Stagecoach Press, 1966), 28; Thomas McFarland, "A Journal of the Coincidences and Acts of Thomas A. McFarland Beginning with the first day of January A.D. 1837," 89, 125, Special Collections, Ralph W. Steen Library, Stephen F. Austin State University, Nacogdoches, Tex.; Finbar Kenneally, O.F.M., *Writings of Fermín Francisco de Lasuén, O.F.M.* (Washington, D.C.: Academy of American Franciscan History, 1965), vol. 2, 336; Eugene Hollon and Ruth Lapham Butler, eds., *William Bollaert's Texas* (Norman: Oklahoma University Press, 1956), 185–56, 377; A. B. Lawrence, *Texas in 1840 or the Emigrant's Guide to the New Republic* (New York: W. W. Allen, 1840), 128; Marcelino Marquinez, letter to Sr. Governor in Santa Cruz, Calif. dated December 13, 1816, San Francisco Chancery Archives.

34. Thomas J. Farnham, "Travels in the Great Western Prairies, the Anahuac and Rocky Mountains and in the Oregon Territory," in Reuben Gold Thwaites, ed., *Travels in the Far Northwest* (Cleveland: A. H. Clark, 1906), vol. 1, 334, 337; Sidney Warren, *Farthest Frontier: The Pacific Northwest* (New York: Macmillan, 1949), 61; Mrs. Juliette Montague Cooke, letter to her cousin Thankful (Smith), dated August 30, 1839, Missionary Letters Collection, Amos S. Cooke and Juliette M. Cooke, folder #5, and Hiram Bingham I, letter to his sisters, dated August 11, 1837, Bingham

Family Papers Collection, Box 1, Hawaiian Mission Children's Society Library, Honolulu, Hawaii.

35. Charles Varlo, *A New System of Husbandry* (Philadelphia: Printed for the Author, 1785), vol. 2, 291; Charles Marshall, *An Introduction to . . . Gardening* (Boston: Etheridge, 1799), vol. 1, 264–65; *Gleanings from the Most Celebrated Books on Husbandry, Gardening and Rural Affairs* (Philadelphia: James Humphreys, 1803), 194.

36. Richard Briggs, *The New Art of Cookery* (Philadelphia: W. Spotswood, R. Campbell, and B. Johnson, 1792), 80; William Woys Weaver, ed., *A Quaker Woman's Cookbook; The Domestic Cookery of Elizabeth Ellicott Lea,* reprint, (Philadelphia: University of Pennsylvania Press, 1982), xxxviii.

37. Thomas Sewall, *A Lecture Delivered at the Opening of the Medical Department of the Columbian College, in the District of Columbia* (Washington: Printed at the Columbian Office, March 30, 1825), 61; Robert Squibb, *The Gardener's Calendar for South-Carolina, Georgia, and North-Carolina: An Eighteenth Century Classic of Southern Gardening,* reprint (Athens: University of Georgia Press, 1980), 59, 89, 106; John Gardiner and David Hepburn, *The American Gardener* (Washington, D.C.: Samuel H. Smith for the Authors, 1804), 27.

38. M'Mahon, *A Catalogue of Garden;* Grant Thorburn, *A Catalogue of Kitchen-Garden, Flower-Seeds, &c* (New York: Southwick and Hardcastle, 1807); Peale, *Memorandum,* vol. 1, back of p. 4.

39. Peale, *Memorandum,* vol. 1, back of p. 4; letter from Thomas Ewing to the Athens County Pioneer Association, dated July 3, 1871, as in "The Autobiography of Thomas Ewing," edited by Clement C. Martzolff, *Ohio Archæological and Historical Quarterly,* 22 (January 1913): 187; *Boston Evening Transcript,* March 31, 1879; *Prairie Farmer,* 50 (June 28, 1879): 202; *Maryland Farmer,* 16 (August 1879): 247; *Reporter and Watch Tower,* August 19, 1885.

40. *Historical Magazine,* 6 (March 1862): 102; *Country Gentleman,* 38 (December 25, 1873): 820; (June 5, 1873): 366.

41. John James Audubon, *The 1826 Journal of John James Audubon Transcribed with an Introduction and Notes by Alice Ford* (Norman, Okla.: University of Oklahoma Press, 1967), 128.

42. Richard J. Hooker, *Food and Drink in America: A History* (Indianapolis & New York: Bobbs-Merrill, 1981), 117–18; *Boston Evening Transcript,* January 20, 1879; Alexander W. Livingston, *Livingston and the Tomato* (Columbus, Ohio: A. W. Livingston's Sons, 1893), 19; Martin Welker, *Farm Life in Central Ohio* (Cleveland: Western Reserve Historical Society Tract No. 86, 1895), vol. 4, 55; letter from Endea Vorer, dated January 29, 1898; *Log*

Cabin Reminiscences, copied by Margaret Parker, compiled by Rev. William Middleswarth (Pomeroy, Ohio: The Meigs County Pioneer and Historical Society, Inc., 1992), 8; Lee Cleveland Corbett and H. P. Gould, "Fruit and Vegetable Productions," *U. S. Department of Agriculture Yearbook, 1925* (Washington D. C.: U.S.D.A., 1925), 415; Samuel M. Welch, *Home History. Recollections of Buffalo* (Buffalo: Peter Paul and Bro., 1891), 76; Allen Baxter, *History of the City of Grand Rapids* (New York & Grand Rapids: Munsell & Company, 1891), 506.

43. *American Farmer*, 4 (May 1822): 40; *Germantown Telegraph*, November 3, 1847; *Country Gentleman*, 19 (May 15, 1862): 318; *Genesee Farmer*, 1 (September 10, 1831): 293; Colin Mackenzie, *Five Thousand Receipts* (Philadelphia: Abraham Small, 1825), 260.

44. *Historical Magazine*, 6 (January 1862): 35–36; *New England Cultivator*, 1 (September 1852): 258; *Lancaster Farmer*, 2 (March 1870): 52.

45. Charles G. Brietz, "Travel Diary," manuscript at Old Salem, Inc., Salem, N.C.; *Florida Agriculturist*, 1 (May 9, 1874): 150.

46. *Genesee Farmer*, 1 (July 30, 1831): 233; *American Farmer*, 14 (September 21, 1832): 222; *Cincinnati Farmer and Mechanic*, 2 (July 30, 1834): 143; Charles F. Crosman, *Gardener's Manual: Culinary Vegetables* (Albany, N.Y.: Hoffman and White, 1835), 21; Eliza Leslie, *The Lady's Receipt-book* (Philadelphia: Carey and Hart, 1847), 44–45; *Godey's Lady's Book*, 61 (July 1860): 74; *Hall's Journal of Health*, 1 (October 1854): 240–43; John Muir, *The Story of My Boyhood and Youth* (Boston & New York: Houghton Mifflin, 1912), 210; Ralph Waldo Emerson, *English Traits* (London: G. Routledge, 1856), 16.

47. *Genesee Farmer*, 5 (March 7, 1835): 78; *Horticulturist*, 4 (September 1849): 422; *Boston Courier*, August 1, 1845.

48. William A. Alcott, *The Young Housekeeper or Thoughts on Food and Cookery* (Boston: George W. Light, 1838), 180; *Maine Farmer*, 6 (July 31, 1838): 193; *Yankee Farmer*, 5 (April 20, 1839): 122–23.

49. Dio Lewis, *Talks about People's Stomach* (Boston: Fields, Osgood, 1870), 147–51.

4

Early Tomato Culture and Cultivation

Troubled by bedbugs, a farmer's wife poured scalding water and turpentine on her bedstead. Unfortunately, after she had enjoyed a short respite from them, the bugs returned. While pondering these circumstances in her garden, she accidentally touched a tomato vine that smelled peculiarly nauseous. This gave her an idea. She immediately rubbed the bedstead thoroughly with the green vine. Her tormentors disappeared and did not return, but she was not sure whether they had died or absconded. She wondered if others had had similar experiences with the tomato plant, so she published her account in the *American Farmer* in 1822.[1] The significance of this anecdote was less the unusual use of the tomato plant and more the source in which it appeared. Launched in Baltimore by John Stuart Skinner, the *American Farmer* was the nation's first highly successful agricultural journal. From its inception in 1819, it published tomato references, articles, and recipes.

Varying in focus, audience, longevity, content, and quality, hundreds of farm journals and periodicals emerged subsequently. These offered practical advice, disseminated information about new farming techniques, provided cookery recipes and domestic advice, announced new varieties of plants, and entertained through fictional stories and poetry. Despite their varied and disparate contents, their influence extended far beyond their subscription bases. During the pre-Civil War period agricultural and horticultural journals provided a national communications and dissemination channel. Subscribers and contributors were often prominent individuals who were active in local agricultural societies and fairs. As most journals were desperate for copy, letters and articles from farmers (and their spouses) readily found their way into print. In turn, other journals and newspapers reprinted them and occasionally offered additional comment.

Periodicals connected American farmers with the latest agricultural thinking in other countries. Five months after the housewife's letter about bedbugs, a surprisingly long article on the tomato was published in the

American Farmer titled "Love-apple. —Solanum; or, Tomato-berry." The article was taken directly from Henry Phillips's *Pomarium Britannicum: An Historical and Botanical Account of Fruits Known in Great Britain.* It reviewed the findings of European herbalists and gardeners, including those of the previously cited Rembert Dodoens, John Gerard, John Parkinson, and Philip Miller. It offered directions for tomato cultivation and proclaimed that tomatoes were abundant in garden markets and were used by all the "great cooks," as they possessed an agreeable acid that was "a very unusual quality in ripe vegetables and which makes it quite distinct from all garden vegetables that are used for culinary purposes in this country." They made good pickles, soups, and sauces, according to this article. They were boiled, roasted, fried with eggs, and preserved in various ways for winter use. Tomatoes were also eaten raw.[2] While the article reflected conditions in England, subsequent American writers echoed similar themes for over thirty years.

Directions for Cultivation

Phillips's directions for tomato cultivation were not the first published in the United States. Directions had been published in the Carolinas in 1774. Robert Squibb published more extensive directions in 1787. He advised gardeners in Georgia and the Carolinas to sow tomatoes in March. In May the plants were to be transplanted "in hills about five feet from each other, and three plants on each hill." In June Squibb recommended sticks for supporting the plants, "which should not be very high, but strong and bushy."[3]

After 1800, directions for tomato cultivation began to be published regularly throughout the United States. In Pennsylvania Bernard M'Mahon recommended sandy soil. He also suggested that tomatoes be planted twelve to fourteen inches apart and watered extensively. In Maryland, John Skinner urged farmers to sow them in a rich soil in early March. When they were a foot high, they should be transplanted three or four feet apart. Six poles wrapped with strings of any kind of bark should surround the plants, he said, to keep them growing up straight.[4]

On Long Island the British gardener William Cobbett declared that tomato seeds "should be sown at a great distance, seeing that the plants occupy a good deal of room." In Ohio the *Farmer and Mechanic* recommended sowing them in early March in hotbeds, "sheltered with grass covers." In Louisiana Thomas Affleck advocated planting them in hotbeds during December and January and transplanting the seedlings in March. In New Hampshire E. M. Tubbs recommended using a hotbed, because they could not be planted in the open until the middle or last of May.[5]

The most extensive pre-Civil War directions for propagating tomatoes were written in 1855 by William Chorlton, a small farmer living on Staten Island. He sowed seeds about the middle of February in a hotbed, upon which he placed "three or four inches of good friable mold." The seeds were covered with a box frame or planted in boxes in a hothouse, where there was a temperature of 50° to 55° at night. When the seedlings were two or three inches high, he transplanted them into other boxes, setting them about six inches apart. When the outside temperature permitted, the plants were lifted with a ball of earth around the roots and transplanted into holes four feet apart and six inches deep. Almost any kind of soil "will answer for the Tomato," he advised. It prospered best, however, and produced fruit of a finer quality with a better flavor, "in a well-drained, tolerably fertile, but not over rich loose mold." A little soil was pressed around the neck and lifted "a little extra up to it, which will encourage fresh roots and strengthen the plant." He suggested sinking poles in an upright position along each row, leaving the tops five feet above the ground about four yards apart. Wires were fastened horizontally to the poles and formed "a cheap trellis to train upon." The branches were then tied loosely to these wires, and a kind of hedgerow was "formed with very little labor." The fruit was "free to the action of air and light."[6] Others suggested wooden trellises, hoops, or fences to support the tomato plant.

Early harvest became the prominent consideration as tomatoes came to be an important cash crop during the 1840s and 1850s. Attention first turned toward forwarding techniques. Cold frames, hotbeds, hothouses, and forcing techniques were recommended. A cold frame consisted of a wooden framework that could be of any length and was usually four feet wide with the back about twenty inches high and the front twelve inches high. Wood crossbars were placed between the sides of the frame, and the outside was staked. The whole was then covered with a glazed sash or a cloth curtain. Cold frames were used only when the outside temperature was greater than 50°.

Hotbeds were similarly constructed, but they had an artificial heating source. The frames were placed over a bed filled with fresh stable manure in a good state of fermentation, which provided heat. The height of manure for a bed depended on the time at which the bed was formed. If it was constructed in February or March, heat was required for two or three months, and about three feet of manure was necessary. But a bed made in April for the purpose of forwarding early plants did not require more than half that quantity. The most common hotbed was composed of several inches of friable mold or manure covered by a few inches of soil. Farmers and gardeners who did not have access to the proper manure heated the hotbed with external heating sources, such as underground brick flues. Hotbeds were positioned so that they faced south to capture the most possible warmth from the sun and were situated to avoid flooding in case of heavy rains.

Hotbeds required constant monitoring. If the heat in the bed was too high, the sash was raised at the back, permitting heat and steam to escape. A mat or old cloth was placed over the opening to keep out the cold wind. In sunny weather the sash was raised. If the weather was very warm, the plants were shaded during the middle of the day. An hour of direct sunshine could destroy a whole bed of plants if the sashes were closed tight. In severe weather, mats or straw were laid over the bed for protection, especially during the night. The bed was kept moist by gentle watering. As the weather became warmer and the plants increased in size, plenty of air was admitted.[7]

The simplest hothouses were little more than large hotbeds heated by means of a stove in the northwest corner. In hot frames and hothouses, as in hotbeds, it was necessary to maintain uniform temperatures and proper circulation of air. Farmers continually opened and closed vents as appropriate. By using these devices and techniques, they expected to produce a crop of tomatoes much earlier than they could by sowing seeds in the open ground. As the earliest tomatoes fetched the most money, the farmer was usually well compensated for the extra trouble of producing the frames and houses.

Other observers suggested procedures designed to produce early tomatoes without the use of hotbeds or similar forwarding accoutrements. A farmer in Palmyra, New York, reported in the *Wayne Sentinel* that plants could be preserved through the winter in a box and set out in the garden in May. The fruit could be picked early in August. The farmer also recommended using tomato cuttings that took root when set out in a hotbed or in open ground after the weather became warm in the spring. In this way, the author opined, "the vegetable may be much advanced, and an early supply obtained with little trouble." John Armstrong, a gardener in New York, suggested sowing seeds in October "in a dry and warm soil, and sheltered situation." The plants were to be covered "with straw, or stable-litter, during the winter." Through the use of this technique, he claimed, the grower could have ripe tomatoes early in the summer.[8] These proposed methods of propagating tomatoes were not successful, however, and were soon disregarded.

Other techniques were proposed to prolong the life of the tomato plant after its normal season ended. Several correspondents suggested that just before the first freeze, farmers should pull up the plant by the roots and hang it upside down in a cool, dark place. There the fruit would continue to ripen, and fresh tomatoes would be available until Christmas.[9]

Today's determinate varieties of tomato plants are relatively short and compact, and their fruit ripens over a brief period of time. Early varieties were indeterminate. The plants were long and straggly, and their fruit continued to set until the frost destroyed the plant. During the 1850s many agricultural journals urged that tomatoes be "bushed" or shortened. The logic was that three-fourths of the fruit was produced upon the part of the vine nearest to the root. Gardeners pinched off the vines beyond a third of the plant's usual growth. This channeled the plant's energy into growing fruit rather than producing unnecessary foliage. According to the editor of the *American Agriculturist,* "the most successful cultivators in our acquaintance made it a rule to let no vine extend beyond four feet from its root." The French mode of growing tomatoes, frequently described in U.S. periodicals during the 1850s, was to pinch off the vine just above the flowers. After the Civil War, gardeners challenged the assumption that shortening tomato plants had any positive impact upon the production of fruit. In 1914 Bert Croft found by chance a seedling that was determinate and self-topping. It was a spontaneous mutation that occurred in a tomato plant in Florida and caused it "to grow in an orderly, compact, determinate fashion." It was called the Cooper Special after C. D. Cooper of Ft. Lauderdale, Florida, who marketed it. This mutation has benefited tomato improvement programs ever since, and most tomatoes available today, except for the heirloom varieties, are determinate.[10]

Edmund Ruffin's *Farmer's Register,* one of the most respected agricultural journals in America, published several notable articles on tomato cultivation during its ten-year life. In an article published in 1841, a correspondent pointed out that, while the tomato was a favorite vegetable on all southern tables, he had never "met with any farmer or gardener who had ever taken the trouble, little as that would be, to ascertain how" tomatoes could be "rendered most productive; or which among our numerous varieties of the latter ripens soonest; or which will yield most; or how many bushels per acre may be calculated on as the average product from ordinary land, of any one variety of the whole."

Many farmers in the southern states relied upon volunteer plants. Plants were set out in vacant spots in gardens and suffered to grow with "little or no culture, and without frames or sticks to support them." Yet everyone knew "perfectly well, that the vines when properly supported, in good garden ground, and planted single, about four or five feet apart, according to the fertility of the ground, will grow to the height of five or six feet, and produce far better than in the common, careless, slovenly way." Country gardeners assigned "to the kitchen garden more ground than enough for their families." In towns, where gardens were small, "the difference between the producer of a well supported and cultivated vine, and one left to spread on the ground," was well worth discovering. A single vine "supported by sticks four or five feet high, or by a frame of the same height, which may easily be made by four or five upright sticks, about 13 inches apart each way, with a twine string passed spirally around them five or six times, and tied at each end, would produce nearly or quite double the quantity that another vine would which was left without support." Six tomato plants, occupying no more than thirty-two square feet, produced an abundance for a whole family.[11]

PESTS AND DISEASES

Agricultural journals were also concerned with tomato pests and diseases. "Black mildew" destroyed the leaves and weakened the blossoms. During their larval stage, tomato borers burrowed into the center of the tomato stalk, where they were impervious to the insecticidal dusts used during the mid-nineteenth century. Red spiders and black flies attacked the tomato fruit. Correspondents in farm journals recommended fumigating plants with tobacco or sulphur dust to eliminate these pests. Other common insects, such as the grasshopper, destroyed whole gardens, including tomato plants, in a matter of minutes. Eliza Farnham, who had established a homestead near San Francisco, had her tomatoes demolished four times in the summer of 1850 by swarms of grasshoppers.[12]

Perhaps the most alarming pest was the large green tomato worm. It was three to four inches long, counting the menacing horn sticking out its back. As it was closely related to the tobacco worm, it must have been a common sight in the southern tobacco-growing areas. In the 1830s northerners began to remark about this pest. In Massachusetts, Ralph Waldo Emerson was concerned that the "young entomologies" were eating up his tomato plants. Others considered the worm "an object of much terror, it being currently regarded as poisonous and imparting a poisonous quality to the fruit if it should chance to crawl upon it."[13]

In New York the worm suddenly appeared during the summer of 1845. A report presented to the New York Farmer's Club stated that it was a very voracious eater: "On a few plants in my garden," the reporter stated, "I found 200 of these worms, large and small." They arrived en masse in Chicago the following year. The *Prairie Farmer* "heard much complaint in this city of the large green worm," which "infested the tomatoes in countless numbers." These worms were described as immense, "often three inches in length and of the size of a man's finger," and were "positively shocking to weak nerves." Some women declared "that unless this nuisance abated, they could not think of venturing among the vines, even if they could persuade themselves to eat tomatoes at all." The writer pointed out that these enormous and "hateful tomato eaters" turned into what at first looked like a hummingbird in the early evening but was really "a very large miller or butterfly, with a tongue or proboscis five or six inches in length—a very beautiful insect." It was called the "Hawk Moth, and belonged to the genus Sphinx, order Lepidoptera." Interest in the tomato worm continued throughout the nineteenth century.[14]

During the 1860s the "killer" tomato worm reappeared. A correspondent from Riker's Hollow in New York asserted that the bite of the tomato worm caused instant death. A girl was said to have been stung by a tomato worm and "and soon after died in terrible agony." Others were "fatally poisoned by the spittle which the worm has power to throw several feet." The *Syracuse Standard* disclosed that a Dr. Fuller had acquired from his garden a tomato worm measuring five inches in length and weighing an ounce. He enclosed it securely in a glass bottle, proclaiming that

> this worm was first discovered this season, and is poisonous as a rattlesnake. It poisons by throwing spittle, which it can throw from one to two feet. This spittle striking the skin, the parts commence at once to swell, and in a few hours death ends the agonies of the patient. Three cases of death in consequence of this poison have been reported. The medical profession is much excited over the new enemy to human existence.

Persons picking tomatoes were advised to wear gloves. Dr. Fuller raised the question about fruit "partly devoured by one of these vermin," might such a tomato not have sufficient virus left upon it to poison the one who eats it? However, Benjamin Walsh, an Illinois entomologist, came to the rescue of the tomato worm, declaring that it was "contrary to all entomological experience, that such harmless creature as the tomato worm should ever do anything of the kind."[15]

Despite this reassurance, an Illinois paper cautioned people to "look out for the worm that infests tomato vines." Its sting was pronounced a "deadly poison." According to the paper, at Red Creek, Wayne County, a servant girl "while gathering tomatoes, received a puncture from one of these worms, which created a sensation similar to that of a wasp sting. In a short time, the poison penetrated to every part of her system, and she was thrown into spasms, which ended in death." The editor of the *Small Fruit Recorder,* who lived not far from Red Creek, concluded that no such event had occurred.[16] His denial only fueled additional talk about the tomato worm. Despite these ominous-sounding stories, the tomato worm is harmless to humans.

THE TOMATO'S ACID CONTENT

The tomato worm was not the only concern raised by commentators with regard to the tomato. Many observers worried about its acid content. At first such concern focused upon the interaction of the tomato's acid with metal or lead glaze in cooking vessels. The Societé d'Horticulture de Paris published an article, frequently reprinted in the United States, stating that the acids in tomatoes "imbibe in the copper vessels, in which they are stewed to a certain consistence, metallic principles, which are injurious to health." Reflecting an awareness of this problem, many recipes called for using earthenware or pewter dishes for cooking and serving tomatoes. John Saxton, the editor of the *Ohio Repository,* recommended putting them on greased plates or pans rather than on common potter's ware as the lead glaze on the latter made it "dangerous with this or any other acid substance." Eli Whitney Blake, a relative of Eli Whitney and also an inventor and scientist, examined the acid contained in tomatoes. He observed that it "acts powerfully on tin, which I believe is not common with the vegetable acids."[17]

The second concern focused upon the effects of the tomato's acid content on the stomach. Most observers acknowledged that the tomato had an "agreeable acid flavor." For instance, the American botanist William Darlington noted that tomatoes were "of a sprightly acid taste." Chester Dewey, a scientist and educator, proclaimed that the "peculiar acid seems to be most grateful to the stomach." The *Cincinnati Farmer and Mechanic* agreed but pointed out that the acid flavor was "not always relished when first tasted." The anxiety with regard to the tomato's acid content may have originated in Britain. The American edition of the British *Cyclopedia or Universal Dictionary* reported that the tomato had a "grateful acidity in Italy, Spain and Portugal," but in England its flavor was insipid and not remarkable "except that few stomachs can bear it in any great quantity." Robley Dunglison, a prominent American physician born in Britain, confessed that "the acid of this vegetable does not agree with every one," but he still thought the tomato wholesome. The New England educator William Alcott believed that some persons were injured by the acid.[18]

During the 1840s, rudimentary chemical analyses of the tomato were published in American periodicals. An analysis conducted in France by F. E. Foderé and E. Hecht revealed that the berries consisted of "a bitter matter, thought to be solinia." The researchers also found a "peculiar acid" in the tomato. J. H. Salisbury of New York found carbonic, silicic, phosphoric, and sulfuric acid. In addition, Salisbury found that more than ninety-four percent of the fruit was water. Of the remainder, tartaric, citric, and malic acids represented three percent and the rest included sugar, potash, magnesium, sodium, chlorine, and other miscellaneous ingredients. These various assessments of the tomato's acid content may have depended upon the variety tested. As the twentieth-century tomato botanists M. Allen Stevens and Charles Rick have pointed out, there is tremendous variation among tomatoes with respect to acidity. Some varieties are highly acidic, while others are extremely mild.[19]

The primary causes of the toxicity of the tomato plant are the alkaloids solanine and demissine and their aglycones. Toxicity is at its highest concentrations in the foliage and stem. Children have developed severe reactions from making "tea" from the leaves of the tomato plant, and cattle and hogs have been poisoned by eating garden cleanings. The green fruit also contains low levels of alkaloids, which become inert as the fruit ripens. Some tomato plants have been bred to produce high levels of tomatine for ointments for treating fungal skin diseases.[20]

UNUSUAL SPECIMENS

Periodicals and newspapers regularly announced unusual specimens of fruits and vegetables. Of particular interest were large tomatoes. In 1822 the *American Farmer* described tomatoes that measured twelve inches in circumference. Two years later it was reported that Timothy Chandler of Caroline County, Virginia, had raised a "tomato of excellent flavour, weighing fourteen ounces, and measuring fourteen inches round." In case the reader should doubt this, the article stated that it had been "weighed and measured by two respectable neighbors." In 1826 a gentleman near Lynchburg, Virginia, raised a tomato two feet and three inches in circumference. The *Norfolk Beacon* reported that Raymond Gervals had one that weighed twenty-seven ounces. The *Gettysburg Star* found one that weighed twenty-eight ounces. The *Knoxville Times* chronicled the appearance of one "weighing *one pound and four* ounces,—and measuring eighteen inches in circumference!" The Richmond *Compiler* "received a present from Mr. John Van Lew, of five prodigious Tomatoes, the aggregate weight of which was 9 1/4 pounds—One of them weighed upwards of *two pounds.*" The Illinois *Sangamo Journal* reported that E. G. Johns of Springfield raised a mammoth tomato that weighted thirty-one and a half ounces. The *Pontiac Jacksonian* stated that J. D. Standish raised one measuring seventeen and a half inches in circumference and weighing thirty ounces. The editor asked, "Can anybody beat that?" Joseph T. Tilden of Newberry, Vermont, did, weighing in with a two-and-a-half-pound tomato, as did the Rev. E. H. Pilcher of Adrian, Michigan, who grew a tomato weighing three pounds. Other monstrosities were noted throughout the 1840s and 1850s. Although no specific measurements were offered, the *Daily Alta California* announced that vegetable stalls on the streets of San Francisco displayed "big luscious tomatoes, larger than the world ever before produced."[21]

There was interest not only in large fruit but also in large tomato plants. The *Botanico-Medical Recorder* reported that the garden of Nicholas Hobson near Nashville produced the "largest tomato plant, probably, that was ever seen." It covered an area of ground that "measured 51 feet in circumference; and had attached to its tendrils the enormous number of *five thousand two hundred and ninety-two tomatoes.*" The *Maine Farmer* reported that Jas Campbell of Camden County grew a tomato vine nine feet high, covering a space of thirty-one square feet and producing 1,500 to 2,000 tomatoes.

Perhaps these reports were exaggerated; perhaps not. The *Guinness Book of World Records* credits Gordon Graham of Edmond, Oklahoma, with growing a plant fifty-three and a half feet long. In 1987 Charles Wilber of Crane Hill, Alabama, grew a single tomato plant that produced 342 pounds of fruit. Another plant produced 16,897 tomatoes in 1988. Gordon Graham is also credited with growing a seven-pound twelve-ounce tomato. New Jersey residents annually vie for $5,000 in cash prizes in the Greater New Jersey Championship Tomato Weigh-In created by Joe Heimbold in 1978. Similar contests are held in Ohio, Pennsylvania, and Georgia. In 1993 Stern's Miracle-Gro Products offered $100,000 to anyone who beat Graham's record. Robert Ambrose has launched the Tomato Club, complete with its own newsletter, to communicate among tomato growers and those trying to beat Graham's and Wilber's records.[22]

Agricultural and Horticultural Societies

Along with the success of farm periodicals was the growth of agricultural and horticultural societies. Societies had existed since the late eighteenth century, but their number mushroomed during the 1830s. Most societies sponsored annual fairs in which the farmers exhibited their stock, fruits, vegetables, and other farm products. Fairs promoted good farming practices, provided a means for communication among farmers, and afforded an occasion for selling commodities and farming equipment. During the 1830s, agricultural societies began offering premiums for tomatoes. In 1833 Conrad Esher received a certificate from the Pennsylvania Horticultural Society "as an honorable testimony of his having gained by his exertion and skill the premium for the growth of the tomatoes submitted to this society this year."[23] By the 1840s most agricultural societies and fairs offered such premiums.

Societies were not interested in the single monstrosities but in multiple tomatoes of standard shape, size, consistency, ripeness, color, and weight. In Connecticut the New Haven Horticultural Society divulged that a premium of $1 was given to Charles Fagen and Eliza Hall for the best half-peck of tomatoes. In New York the Delaware County Fair and Cattle Show gave $1 "for best 10 Tomatoes." In Ohio the Montgomery County Horticultural Society awarded a 50-cent premium to A. Fowler "for the best peck of tomatoes." During the 1850s, premiums offered by agricultural societies dramatically increased in all regions of the nation. Horticultural societies gave them for tomato products such as ketchup and pickles. By the mid-1850s, premiums were offered for canning and preserving tomatoes. In Tennessee, agricultural fairs offered premiums for tomatoes "put up in air-tight cans."[24]

SEEDS, VARIETIES, AND FRESH TOMATOES

In the 1790s Cuthbert and David Landreth sold fruits and vegetables from a garden stall by the side of the old Philadelphia courthouse. According to the historian James Boyd, the Landreths sold tomatoes to the French immigrants, but there was little demand from others. Tomato seeds were sold in Philadelphia by 1800. John Lithen's broadside noted "love apple" seeds for sale about the same time as Bernard M'Mahon's broadside advertised them. In New York tomato seeds were sold in 1807 by Grant Thorburn, who established a seed farm and began selling seeds shortly after the turn of the century. Seeds were sold in Baltimore by 1810 and in Boston by 1827. David Landreth Jr. opened a store in Charleston, which was then the only seed store in the southern United States. The store flourished until the Civil War. During the 1820s and 1830s the Landreths began franchising the sale of their seeds in other states. By the 1830s tomato seeds were sold throughout the country.[25]

Fresh tomatoes were regularly sold in markets during the 1820s and 1830s. The prices in Washington Market in New York City demonstrated a common pattern. In July tomatoes sold for a dollar and fifty cents a peck, which is about eight quarts. By August they were down to twenty-five cents a bushel, which consists of four pecks. By October the price had risen to fifty cents a bushel. Prices at Boston's Quincy Market followed the same pattern. For instance, in 1836, prices went from seventy-five cents a dozen in July to twenty-five cents a peck in September to fifty cents a peck in November. By 1845 tomatoes sold for a dollar a dozen in July and dropped to seventeen cents a peck in August. A similar pattern was evident in Baltimore:

> On [the tomato's] first appearance in our market this year, it sold for 25 cents per dozen, which was certainly very high for times of unusual plenty. Now the same quality may be had, of the finest size and flavor, for 25 cents per peck. That price places them within the reach of the community at large. But they will be unusually abundant, and will doubtless descend to rates which will scarcely justify the gathering.[26]

In 1834 the editor of the *Cincinnati Mirror* complained that in mid-July the cost of tomatoes was fifty cents a dozen. The complaint generated a response from the editor of the Cincinnati *Farmer and Mechanic,* who conjectured that the cost in July was high because farmers in the Cincinnati area failed to use forcing techniques. Those who did were rewarded for their efforts.[27]

This pattern continued throughout the pre-Civil War period. Edmund Morris, who began farming in 1855 on a ten-acre farm in southern New Jersey, shipped tomatoes to markets in both Philadelphia and New York City. The first tomatoes that he sent to market, though not perfectly ripe, sold "for sixpence apiece." He managed to get twenty baskets into the New York market among the very first of the season, where they netted him sixty dollars. The next twenty bushels netted twenty-five dollars, the next fifty only fifteen dollars.

> After that the usual glut came on, and down went the price to twenty and even fifteen cents. But at twenty and twenty-five I continued to forward to Philadelphia, where they paid better than to let them rot on the ground. From 200 baskets at these low prices I netted $35. Then, in the height of the season, all picking was suspended, except for pigs, who thus had any quantity they could consume. But the glut gradually subsided as tomatoes perished on the vines, and the price again rose in market to twenty-five cents, then to fifty, then to a dollar, and upwards. But my single acre afforded me but few at the close of the season. I did not manage to realize $40 from the fag-end of the year, making a total net yield of $190.[28]

The pattern was simple. So was the message: early tomatoes generated a handsome profit, while tomatoes maturing at the height of the season often rotted on the vine. Farmers were encouraged to save the seed of the earliest tomatoes with the expectation that they would produce even earlier fruit the following year. In 1824 John Skinner reported that they needed to be planted from seed every year and not "suffered to come up spontaneously." He suggested that farmers chose one fruit while it was growing on the vine and "dry the seed in the shade." It also became evident that tomatoes with certain qualities were more salable than others. William Chorlton urged everyone who had a tomato plot to notice the plants when they were in full bearing: "One or more will show more excellence than the other—pick from the very best, the most desirable fruit, and save them for seed. Repeat this each season, always having an eye to form, color, productiveness, flavor, and size." As the average tomato fruit contained from 250 to 300 seeds, farmers and gardeners needed to save only a few tomatoes to guarantee enough seed for the following year. The *Indiana Farmer* offered the following instructions on how to clean tomato seeds:

> Before fermentation has commenced in the tomato, place in a strong cloth or sack, of as loose a texture as the retention of the seeds will admit of, a convenient quantity of seeds and pulp; after pressing out as much of the latter as will easily pass through, frequently dip the cloth or sack in water, then rub and squeeze it till the seeds are clean enough to be finished by rinsing.[29]

In Rochester, J. Slater began saving seeds from the roundest and smoothest tomatoes he could find. His tomatoes were neither flat nor wrinkled, "but as round as an orange, and as smooth and large as the largest Northern Spy apple." Dr. T. J. Hand, originally from Sing Sing, New York, began crossing the small cherry tomato with several larger lumpy varieties. His efforts were rewarded when he produced a tomato with a solid mass of flesh and juice, small seeds, and smooth skin. Under the name Trophy tomato, its success was "unbounded" after the Civil War, with its promotor, Colonel George Waring, selling seeds for twenty-five cents apiece.[30]

Tomato varieties were rarely listed in early broadsides and seed catalogues, but there were many different types of tomatoes grown in the United States. Thomas Jefferson exported tomato seeds from France during the early 1780s. In 1809 General John Mason sent him "Spanish tomato" seeds whose fruit was "very much larger than common kinds." Jefferson also planted what he called dwarf tomatoes, which may have been cherry tomatoes. In 1824 he imported seeds from Mexico. During the 1820s large and small varieties with red and yellow fruit were noted. By the 1830s this spiraled to several types, differing in size, shape, and color. In 1835 self-proclaimed botanist and medical practitioner Constantine Rafinesque enumerated fourteen varieties, although from the descriptions that have survived it is difficult to determine the distinctions among many of them. In 1840 the *Genesee Farmer* advertised large red, large yellow, small red cherry, and Cuba or Spanish tomatoes. More varieties appeared as the decade progressed, such as pear-shaped, cluster, preserving, fig-shaped, yellow cherry, and egg-shaped varieties.[31]

Of special interest was a variety brought back from the Pacific in 1841. The *U.S. Gazette* reported that the American Exploring Expedition in the South Pacific had run across one called the Fegee Tomato. A sailor had sent seeds back to a friend in Philadelphia, and Charles Wilkes, the captain of the expedition, sent seeds to the secretary of the navy, James Pauling, who evidently dispersed them in the United States. They had no discernable effect upon tomato culture and died out after a few years of cultivation. This variety was later incorrectly identified as the forerunner of the actual Fejee Tomato, which became a popular variety after the Civil War. Despite its name, this variety originated in Italy.[32]

Several plants advertized as tomato varieties were not botanically related to *Lycopersicon esculentum* at all, including the Tree Tomato and the Cape Gooseberry. This suggests that by the 1850s the name *tomato* was in such high esteem that it was used to sell other plants. Regardless of the inflation in the number of purported tomato types, the *American Agriculturist* maintained that four varieties were most esteemed and cultivated. The *large smooth-skinned red,* an excellent variety, differed "from all other large sorts, in having a smooth skin entirely free from protuberances or inequalities of any kind." The *common large red,* "with the fruit depressed at both ends, furrowed on the sides, and varying in circumference, from three to eighteen inches," was a prolific bearer and was "universally cultivated." The *pear-shaped* was much smaller than either of the preceding, very fleshy, and contained fewer seeds. The *cherry-shaped red* was a "beautiful little fruit, much resembling a cherry in size and appearance." While some tomatoes were considered oddities or curiosities, there was a nascent relationship between tomato varieties and their culinary usage. Red tomatoes were best for ketchup and cooking. Fig-shaped tomatoes were frequently recommended for making confections. Pear-shaped, cherry-shaped, yellow types, and the pink-red tomatoes were for pickling.[33]

Farmers and gardeners slowly bred tomatoes with particular characteristics in the tomato fruit such as a round shape; a smooth-skin; solid flesh; and ripeness all over. William Chorlton believed that if agricultural societies and fairs rewarded these characteristics, the standards for tomato quality would rise, and the large growers would soon be "forced to take a better sample to the city, instead of the thick skinned, hollow subjects, which are too often seen on the huckster's stall, which 'bounce' like a foot-ball."[34]

The benefits of these breeding efforts first became apparent just as the Civil War began. Fearing Burr's *Field and Garden Vegetables of America,* initially published in 1863, reflected his experience as a seedsman and gardener in Massachusetts during the late 1850s. He listed twenty-two tomato varieties, only one of which was not botanically within the *Lycopersicon* genus. His 1865 edition listed two additional varieties.[35]

Precisely what these varieties looked like is unknown. Few illustrations of specific varieties have been located, with the major exception of the illustrations in Fearing Burr's book and a few illustrations in agricultural journals. Vegetables were never a popular subject for still-life artists, and only four American paintings containing tomatoes are known to have survived. These paintings, however—one by Raphaelle Peale dated to about 1795, one by an unknown artist painted in the 1830s, and two by Paul Lacroix painted in 1863 and 1865—show dramatic changes in the tomato's shape. Peale's tomato is extremely ribbed and lumpy; the next is less lumpy, but extremely large; and Lacroix's tomatoes are much smoother and more closely resemble today's varieties. The color of all the tomatoes is bright red. Virginia Granbery did a painting called *Yellow Tomato,* but its date and whereabouts are unknown.[36]

A THOUSAND AND ONE USES

Beginning in 1822, agricultural periodicals and newspapers began publishing hundreds of tomato recipes. Tomatoes were applied to "a thousand and one uses," many of which were for nonculinary purposes. The *New York Courier & Enquirer* posited that they made green dye and imparted a "beautiful orange color to animal oils." The *Wheeling Times,* stating that the tomato removed strains of ink, and what was "commonly called 'iron rust,' from linen," pronounced it "a very useful vegetable." The editors of the *Illinois State Register* believed that its unadulterated juice completely removed "all stains of fruit, &c. and marks of *iron mould*, from all linens and muslins." Nothing more was necessary than "its mere application, and exposure to the sun until dry." Its simplicity and efficacy rendered it worth remembering.[37]

Others recommended tomatoes for curing scours in pigs; deterring pests; and feeding domestic fowl, hogs, and cattle. South Carolina's *Cheraw Gazette* claimed that, when cows ate tomatoes, the quantity of milk increased and the butter yielded a finer flavor. They did not doubt that tomatoes produced "*more* food for cattle, and of *better* quality," and were more easily raised on a "given portion of ground, and at *less* expense" than any other food product. Virginia's *Farmer's Register* claimed that cows ate "the vine with great avidity." New York's *American Agriculturist* recommended frying and boiling tomatoes along with pumpkins and other vegetables before serving them to cows. A correspondent in Missouri's *Valley Farmer* boiled tomatoes with squash and gave this mix to his cows daily. The most striking result "was the beautiful yellow color, and the delicious flavor imparted to the butter by the tomatoes."[38]

NOTES

1. *American Farmer,* 4 (May 3, 1822): 40.

2. *American Farmer,* 4 (October 4, 1822): 218–19; Henry Phillips, *Pomarium Britannicum: An Historical and Botanical Account of Fruits known in Great Britain* (London: Printed for the Author, 1820), 225–27.

3. Robert Squibb, *The Gardener's Calendar for South-Carolina, Georgia, and North-Carolina: An Eighteenth Century Classic of Southern Gardening* (Athens: University of Georgia Press, 1980), 59, 89, 106.

4. Bernard M'Mahon, *The American Gardener's Calendar; Adapted to the Climates and Seasons of the United States* (Philadelphia: Printed by B. Graves for the Author, 1806), 200, 319, 372, 401, 429; *American Farmer,* 6 (August 27, 1824): 183.

5. William Cobbett, *The American Gardener* (Claremont, N.H.: Manufacturing Company, 1819), 160–61; *American Farmer,* 4 (February 1823): 371; *New England Farmer,* 6 (May 16, 1828): 339; *Cincinnati Farmer and Mechanic,* 2 (July 30, 1834): 143; Thomas Affleck, *Southern Rural Almanac and Plantation and Garden Calendar for 1851* (New Orleans: J. C. Morgan, 1850), 21, 26, 32, 38, 45, 74; E. M. Tubbs, *The New Hampshire Kitchen, Fruit, and Floral Gardener* (Peterboro, N.H.: K. C. Scott, 1852), 50–52.

6. *Horticulturist,* 10 (March 1, 1855): 130–34.

7. *Q. C. Almanac,* as in the *New England Farmer,* 1 (March 17, 1849): 107.

8. John Armstrong, "Treatise on Gardening," *Memoirs of New-York Board of Agriculture* (Albany, N.Y.: Packard & Van Benthuysen, 1823), vol. 2, 89; *New York Farmer,* 4 (June 1831): 152; *American Farmer,* 3d ser., 3 (March 1842): 324; 15 (February 1834): 407; *Wayne Sentinel,* as reported in the *Cultivator,* 9 (October 1842): 165; *Working Farmer,* 3 (March 1, 1851): 15.

9. *American Agriculturist,* 7 (May 1848): 137–38; *Farmer and Mechanic,* 2d ser., 4 (January 24, 1850): 46.

10. *American Agriculturist,* 15 (August 1856): 256; Charles Rick, "The Tomato," *Scientific American,* 239 (August 1978): 80; Victor R. Boswell, "Improvement and Genetics of Tomatoes, Peppers, and Eggplant," U. S.

Department of Agriculture's *Yearbook, 1937* (Washington, D.C.: U.S.D.A., 1938), 181.

11. *Farmer's Register,* 9 (September 1841): 589.

12. *Northern Farmer,* 1 (June 1854): 66–67; *Horticulturist,* 10 (March 1, 1855): 130–34; Eliza W. Farnham, *California: In-down and Out* (New York: Dix, Edwards, 1856), 132.

13. Letter from Emerson Ralph Waldo to Margaret Fuller, dated June 28, 1838; *Illustrated Annual of Rural Affairs,* 1869, pp. 205–7.

14. *New York Farmer,* 3 (November 1845): 387; *Prairie Farmer,* 7 (June 1847): 179; *Country Gentleman,* 9 (March 12, 1857): 176.

15. *Country Gentleman,* 28 (August 23, 1866): 126; *Illustrated Annual of Rural Affairs,* 1869, pp. 205–7; *Ohio Farmer,* 18 (October 23, 1869): 680.

16. *Small Fruit Recorder,* 3 (October 1871): 157.

17. Annales de la Societe d'Horticulture de Paris, as published in the *New York Farmer,* 4 (August 1831): 208; *Ohio Repository,* August 21, 1845; *Massachusetts Ploughman,* 4 (September 20, 1845); Eli W. Blake, "Acid in Tomatos," *American Journal of Science and Arts,* 17 (January 1830): 115–16.

18. John B. Russell, *Catalogue of Kitchen Garden, Herb, Tree, Field and Flower Seeds* (Boston: New England Farmer Office, 1827), 15; George Barrett, *Catalogue of Kitchen Garden, Herb, Tree, Flower and Grass Seeds* (Boston: J. Ford, 1833), 18; Thomas Bridgeman, *The Young Gardener's Assistant* (New York: Booth & Smith, 1833), 69; William Darlington, *Florula Cestrica* (West-Chester, Pa.: Printed for the Author by Simon Siegfried, 1826), 117; Chester Dewey, *Report on the Herbaceous Plants and on the Quadrupeds of Massachusetts* (Cambridge: Folsom, Wells, and Thurston, 1840), 166; *Cincinnati Farmer and Mechanic,* 2 (July 30, 1834): 143; *Botanico-Medical Recorder,* 6 (April 7, 1838): 217–18; Abraham Reese, *The Cyclopedia or Universal Dictionary of Arts, Sciences, and Architecture* (Philadelphia: Samuel F. Bradford and Murray, Fairman, 1825), vol. 22; Robley Dunglison, *Elements of Health* (Philadelphia: Carey, Lea & Blanchard, 1835), 300; William A. Alcott, *The Young Housekeeper or Thoughts on Food and Cookery* (Boston: George W. Light, 1838), 180.

19. M. Allen Stevens and C. M. Rick, "Genetics and Breeding," in J. G. Atherton and J. Rudich, eds., *The Tomato Crop* (New York: Chapman and Hall, 1986), 86; F. E. Foderé and E. Hecht, "Analyse des feuilles et des fruits de la pomme d'amour (solanum lycopersicum)," *Journal de Pharmacie des sciences accessoires,* 2d ser., 18 (1832): 105–12; J. H. Salisbury, *Transactions of the New York State Agricultural Society,* 8 (1848): 370–73.

20. Randy G. Westbrooks and James W. Preacher, *Poisonous Plants of Eastern North America* (Columbia: University of South Carolina Press, 1986),

155; James W. Hardin and Jay M. Arena, *Human Poisoning from Native and Cultivated Plants*, 2d edition (Durham, N.C.: Duke University Press, 1974), 140; J. M. Kingsbury, *Poisonous Plants of the United States and Canada* (Englewood Cliffs, N.J.: Prentice-Hall, 1964), 284; Charles Rick, "The Tomato," *Scientific American,* 239 (August 1978): 80.

21. *American Farmer,* 4 (October 1822): 218–19; 6 (August 27, 1824): 183; 8 (November 1826): 261; Joseph Sabine, "On the Love Apple or Tomato, and an Account of its Cultivation; with a Description of Several Varieties, and some Observations on the Different Species of the Genus Lycopersicum," *Royal Horticultural Society Transactions,* 3 (January 1819): 342–54; *Norfolk Beacon,* as in the *Farmer and Gardener,* 3d ser., 2 (July 28, 1835): 97; *Gettysburg Star,* as in the *Constitution, Farmers' and Mechanics' Advertiser,* September 15, 1835; *Knoxville Times,* as in the *New York Evening Star,* August 7, 1839; *Richmond Compiler,* as in the *New Haven Palladium,* August 10, 1839; *Sangamo Journal,* September 9, 1842; *Pontiac Jacksonian,* as in the *Michigan Farmer,* 1 (September 15, 1843): 117; 12 (February 1854): 50; *Western Farmer,* 4 (October 1843): 68; (December 1843): 116; *Central New York Farmer,* 3 (September 1843): 292; *New England Farmer,* 2 (October 19, 1850): 344; *Daily Alta California,* September 13, 1851.

22. *Botanico-Medical Recorder,* 6 (April 7, 1838): 217–18; *Maine Farmer,* 22 (September 21, 1854); *Guinness Book of World Records* (New York: Facts on File, 1992), 48; *Guinness Book of World Records* (New York: Sterling, 1987), 76; "$100,000 Tomato Official Rules," (Port Washington, N.Y.: Stern's Miracle-Gro Products, n.d.); *Tomato Club,* 6 (August 1993).

23. *Meehans' Monthly,* 6 (March 1896): 58; *American Farmer,* 2d ser., 3 (September 13, 1836): 158.

24. *Connecticut Farmer's Gazette and Horticultural Repository,* 1 (September 18, 1840): 13; *United States Farmer,* 1 (July 1842): 221; *Central New York Farmer,* 2 (June 1843): 9; 3 (February 1843): 63; *Connecticut Farmer,* 2d ser., 1 (December 15, 1844): 54; 4 (October 1, 1844): 357; *Vermont Family Visitor,* 1 (April 1846): 325; *Genesee Farmer,* 8 (May 1847): 119; *Magazine of Horticulture,* 14 (July 1848): 331; Delaware County Agricultural Society, 1847; *Proceedings of the Montgomery County Horticultural Society* (September 18, 1847): 33; E. G. Eastman, *Biennial Report of the State Agricultural Bureau of Tennessee to the Legislature of 1855–6* (Nashville: G. C. Torbett, 1856), 185.

25. James Boyd, *A History of the Pennsylvania Horticultural Society, 1827–1927* (Philadelphia: Printed for the Society, 1929), 22–23; A. J. Pieters, "Seed Selling, Seed Growing, and Seed Testing," *Yearbook of the United States Department of Agriculture* (Washington, D.C.: U.S.D.A., 1900), 552; John Lithen, "Catalogue of Garden Seeds" (Philadelphia, c.1800); M'Mahon,

A Catalogue of Garden, Grass, Herb, Flower, Roots (Philadelphia, c.1800); Thorburn, *A Catalogue;* William Booth, *A Catalogue of Kitchen Garden Seeds and Plants* (Baltimore: G. Dobbin and Murphy, 1810), 4; Russell, *Catalogue,* 15; *New Orleans Bee,* January 4, 1832; *Mississippi Free Trader & Natchez Gazette,* 1 (March 18, 1836), 131; Frank Wiling Leach, *Landreth Family; One of a Series of Sketches Written by Frank Wiling Leach for the Philadelphia North American, 1907–1913, and Brought Down to Date* (Philadelphia: Historical Publication Society, 1934).

26. *New York Farmer,* 2 (July 1829): 176; (August 1829): 200; (October 1829): 252; *Botanico-Medical Recorder,* 6 (April 7, 1838): 217–18; *Magazine of Horticulture,* 2 (August 1836): 316; (October 1836): 398; (November 1836): 438; 7 (August 1845): 318; (September 1845): 358; *Baltimore Sun,* August 3, 1839.

27. *Cincinnati Mirror,* 3 (July 19, 1834): 319; *Cincinnati Farmer and Mechanic,* 2 (July 30, 1834): 143.

28. Edmund Morris, *Ten Acres Enough* (New York: J. Miller, 1864), 118, 156–57.

29. *American Farmer,* 6 (August 27, 1824): 183; *Horticulturist,* 10 (March 1, 1855): 130–34; *Indiana Farmer,* 9 (October 15, 1860): 315.

30. *Genesee Farmer,* 17 (May 1856): 156; Liberty Hyde Bailey, *The Survival of the Unlike* (New York: Macmillan, 1897), 481; U. P. Hedrick, ed., *Sturtevant's Edible Plants of the World* (New York: Dover, 1972), 460.

31. Thomas Jefferson, *Thomas Jefferson's Garden Book 1766–1824 with Relevant Extracts from his other Writings,* annotated by Edwin Morris Betts (Philadelphia: American Philosophical Society, 1981), 403, 564, 613; *New England Farmer,* 10 (March 1832): 285; *Botanico-Medical Recorder,* 6 (November 18, 1837): 58–59; *Genesee Farmer,* 2d ser., 1 (February 1840): 32; Dunlap & Thomsom, *Catalogue of Seeds, Roots, and Shrubs* (New York: J. W. Bell, 1847), 6; Warren's Garden and Nursery, *Annual Descriptive Catalogue of Fruit and Ornamental Trees* (Boston: 1845–56), 63.

32. *U.S. Gazette,* as in the *American Farmer,* 2d ser., 3 (October 1841): 181; Charles Wilkes, *Narrative of the U.S. Exploring Expedition* (London: Wiley and Putnam, 1845), vol. 3, 309, 335; Nelson Klose, *America's Crop Heritage: The History of Foreign Plant Introduction by the Federal Government* (Ames, Iowa: Iowa State College Press, 1950), 29.

33. *American Agriculturist,* 7 (May 1848): 137–38; *Genesee Farmer,* 1 (July 1831): 233.

34. *Horticulturist,* 10 (March 1, 1855): 130–34.

35. Fearing Burr, *The Field and Garden Vegetables of America* (Boston: Crosby and Nichols, 1863), 643–54; (Boston: J. E. Tilton, 1865), 628–42.

36. *Still Life with Vegetables and Fruits,* painting by Raphaelle Peale at the Wadsworth Atheneum in Hartford, Conn.; the second painting is untitled and is in a private collection; William Gerdts, *American Still-Life Painting* (New York: Praeger, 1971), plate VIII. The location of the 1863 painting by Lacroix is unknown, but a photograph of it is in the possession of William Gerdts. Other paintings were exhibited prior to 1876, but their whereabouts are unknown. These include two paintings by Virginia Granbery.

37. *American Farmer,* 3d ser., 3 (May 1842): 414; *New York Courier & Enquirer,* September 17, 1835; *Wheeling Times,* as in the *Liberty Hall and Cincinnati Gazette,* August 29, 1837; *Illinois State Register,* September 14, 1839.

38. *American Farmer,* 4 (May 1822): 40; 3d ser., 3 (June 1841): 41; 3d ser., 5 (September 1843): 135; *Farmer's Register,* 7 (August 1839): 495; 9 (September 1841): 589; *New England Farmer,* 10 (October 1832): 101; 21 (September 1842): 81; *Cheraw Gazette,* as published in the *Southern Planter,* 3 (August 1843): 173–74; *American Agriculturist,* 15 (July 1856): 228; *Valley Farmer,* 9 (January 1857): 2.

5

Early American Tomato Cookery

British cookbooks were published in America beginning in 1742, but they did not accurately reflect American cookery. To be sure, British cookery practices dominated the colonial kitchen, but physical isolation, different climatic conditions, and New World foods contributed to culinary drift. The cookery of other European refugees and colonists, slaves from Africa and the Caribbean, and Native Americans mingled and melded with the preeminent English style, creating a unique melting pot for American cookery.

Eighteenth-century cookbooks did not adequately mirror the advances of tomato cookery in Britain or the British North American colonies; only one recipe with tomatoes as an ingredient was published in the English language prior to 1804. This was of Spanish origin and was published initially in a supplement to Hannah Glasse's *The Art of Cookery* and later revised by Richard Briggs in *The New Art of Cookery.* In 1804 Dr. Alexander Hunter, a Scottish physician who was convinced that no one could "be a good physician who has not a competent knowledge of cookery," published a cookbook under the imposing title *Culina Famulatrix Medicinae; or, Receipts on Modern Cookery.* Hunter had studied medicine in France and had acquired many of his recipes from Continental European sources. According to him, tomatoes had a pleasant acid taste and were much used by the Spaniards and Portuguese. They made "a charming sauce for all kinds of meat, whether hot or cold."

Hunter's cookbook contained two recipes for tomato sauce and one for "potting" or preserving tomatoes. His first tomato sauce recipe called for stewing the tomatoes until they were soft, removing the skins, and seasoning with garlic, ginger, salt, and chili vinegar or cayenne pepper. In his second recipe he added thinly sliced shallots, which were removed before bottling. The presence of chili vinegar or cayenne pepper made it a spicy mixture. While both recipes were bottled, they were intended for use within a short period of time. For potting tomatoes, Hunter stuffed tomato pulp into small stone pots, pouring "over the surface some melted mutton suet" and tying a wet bladder over the top to keep out the air. He recommended using smaller pots for storage, "as the pulp spoils after being opened."

Since tomatoes were often difficult to obtain in Britain, Hunter recommended that cooks raise their own in hothouses. He also included a recipe for "Mock Tomata Sauce" made from "sharp-tasted apples." This recipe suggested that by the turn of the nineteenth century there were people who liked the flavor of the tomato enough to attempt to duplicate its taste if they did not have access to tomatoes.[1]

Hunter's recipes were published regularly by other cookbook authors for over fifty years. In 1807 British cookbook author Maria Eliza Rundell purloined one of them for tomato sauce, retitling it "Tomata Sauce, for hot and cold Meats." It was published in America seven years later. In 1821 American physician Thomas Cooper added a glass of Madeira to improve the recipe. The following year a British physician, William Kitchiner, published Hunter's recipes for "Tomato, or Love-apple Sauce" and for "Mock Tomato Sauce."[2]

In America tomato sauce recipes were included in cookery manuscripts during the early nineteenth century. A Mrs. Read, from New Castle, Delaware, compiled a recipe book that contained recipes for tomato sauce, tomatoes as a vegetable, and tomato and okra soup. Her recipe for "Tomatoe Sauce" included onions, salt, pepper, allspice, cloves, and mace. This recipe instructed the cook to strain out the skin and seeds and boil the concoction to the consistency of a jelly. A similar recipe was later published in the *American Farmer* but was retitled "Tomata Catsup."[3]

Tomato sauce recipes were published regularly in Britain. Louis Eustache Ude, the "ci-devant cook to Louis XVI" who had wisely removed to England during the French Revolution, published a complex recipe for "Sauce aux Tomates," translated into English as "Love-Apples Sauce." Besides tomatoes, it included onions, bits of ham, a clove, some thyme, and *Espagnole* sauce. The mixture was to be rubbed through a tammy cloth and cooked for twenty minutes and the resulting purée to be served immediately upon completion. This recipe presupposed, of course, that the cook already possessed Espagnole sauce, which contained ham, veal, mushrooms, parsley, and green onions. Espagnole sauce also required some consommé, which was made with either veal, other meat or fowl, and coulis, which in turn required brown roux with veal gravy. Brown roux was made with butter and flour, which, thankfully, could be kept for a long time. Ude employed his "Love-Apples Sauce" as an ingredient in recipes for calf's head, calf's brain, calf's ears, calf's feet, tendons of veal, legs of fowl, chicken, rabbit, pigeon, and other such treats. His cookbook, *The French Cook,* was published in Britain in 1813 and in America fifteen years later. His "Love-Apples Sauce" recipe was frequently expropriated in Britain and America by other cookbook authors.[4]

While chefs in restaurants might make such a complicated sauce, the average cook was unlikely to go through the necessary gyrations. Simpler recipes for tomato sauce soon appeared. In 1814 *The Universal Receipt Book,* attributed to Richard Alsop, included the first known American recipe for "Tomato or Love Apple Sauce." This version consisted of little more than stewed tomatoes seasoned with salt and pepper and diluted with water.[5] Beginning in the 1820s, numerous tomato sauce recipes with but slight variations were published in America.

MARY RANDOLPH

Amelia Simmons published the first American cookbook in 1796. It mainly reflected English cookery practices and contained no tomato recipes. While several tomato recipes had been published in America prior to 1824, most had been previously published in Great Britain. As culinary historian Karen Hess has pointed out, Mary Randolph's *Virginia House-wife* made a break from British cookery practices and a breakthrough for tomato cookery in America. Based upon Randolph's experiences managing a boardinghouse, the first and second editions of her cookbook, published in 1824 and 1825, infused tomatoes into seventeen recipes, including one for ketchup, two for marmalade, three for soup, and four for Spanish dishes.[6]

Randolph's contribution was not just in the quantity of her tomato recipes. She demonstrated the wide range of the tomato's culinary usage from soup to sweets and from breakfast to dinner dishes. While some of her recipes can be traced to other sources, all demonstrated modifications based upon extensive culinary experience. Her tomato recipes were influential in American cookery. Many were subsequently copied and revised, such as her recipes for "scolloped" (or baked) tomato, tomato soy, and tomato marmalade.

As pivotal as Mary Randolph's contributions were, American tomato cookery did not just blossom forth from her unique efforts. In part her cookbook was based upon tomato recipes developed and published in Britain and France. In part it reflected unrecorded developments that occurred previously over decades, particularly in Virginia. And in part it demonstrated her willingness to absorb and refine recipes from other cultures, as illustrated by her inclusion of those of Spanish origin. Her recipes were more than just a manifestation of the state of tomato cookery in America or her ability to borrow and enhance recipes. She set the standards for tomato cookery for the next three decades, and subsequent cookbook authors such as

N. K. M. Lee, Eliza Leslie, Lettice Bryan, and Sarah Rutledge borrowed extensively from her work, as did newspapers, magazines, and agricultural publications.

LYDIA MARIA CHILD AND N. K. M. LEE

From 1825, American cookbooks regularly included at least a few tomato recipes. Massachusetts-born Lydia Maria Child published *The Frugal Housewife* in 1829. It summarized the lessons she learned in running a low-budget household in New England. It included recipes for stewed tomatoes and tomato ketchup. She averred that "the best sort of catsup is made from tomatoes" and that a cup of ketchup added to chowder made the dish "very excellent." This was the first known inclusion of tomatoes in chowder. Ironically, a variation of this recipe later became known as Manhattan clam chowder, even though Child's recipe was first published in Boston. Later editions of her cookbook included the first known recipe for tomato pie, which was similar to "rich squash pie" with the addition of an extra egg or two.[7]

N. K. M. Lee's *The Cook's Own Book,* first published in 1832, was essentially a compendium of recipes compiled from British and American sources, including twelve tomato recipes. Four were for tomato sauce, one of which was Ude's. Her "French" tomato sauce recipe required stewing tomatoes and onions for forty-five minutes and seasoning them with parsley, thyme, a clove, and butter. This mixture was to be strained through a horsehair sieve and served immediately. Her "Italian" recipe contained the same ingredients as the French recipe with the addition of a bay leaf, allspice, and saffron. The sauce was as thick as a purée. In her final recipe for tomato sauce she added vinegar to baked tomatoes and seasoned them with cayenne pepper. This sauce was as thick as cream.

Lee's tomato soup recipe was more of a vegetable soup; it contained three large carrots, three heads of celery, four large onions, two large turnips, half a pound of lean ham, three quarts of brown gravy soup, and only eight to ten tomatoes. The stewed concoction was sieved and served with diced fried bread. Her two recipes for preserving tomatoes basically provided directions for storing whole tomatoes in a brine. She claimed that, preserved in this way, tomatoes would survive for a year, but they had to be soaked for several hours in fresh water before they could be used. Her three tomato ketchup recipes were similar to those previously published, and her "Tomata Marmalade" recipe derived from Randolph.

Though most of Lee's recipes were borrowed from other sources and were not necessarily reflective of culinary practices in New England, the number of her tomato recipes signified the growing acceptance of the tomato even in that area during the early 1830s. While some New Englanders may have shunned the tomato, Child's and Lee's recipes clearly prove that others were seriously experimenting with it. The popularity of the cookbooks by Child and Lee was demonstrated by the dozens of editions they went through and by the frequency with which their recipes were borrowed by others.[8]

ELIZA LESLIE

No pre-Civil War cookbook author with the exception of Mary Randolph made a greater contribution to tomato cookery in America than did Philadelphia-born Eliza Leslie. Leslie attended Elizabeth Goodfellow's cooking school. In 1834 she published recipes that she had learned there in *Seventy-Five Receipts for Pastry, Cakes, and Sweetmeats.* Subsequent editions of this work included recipes for tomato "sweetmeat" and ketchup. Leslie's *Domestic French Cookery,* published in 1832, was in part translated recipes from the 1829 edition of Louis Eustache Audot's *La Cuisinière de la ville.* It included five tomato recipes. Her *Directions for Cookery,* originally published in 1837, contained thirteen recipes with tomatoes as an ingredient. Later editions added fifteen more. In 1843 *Miss Leslie's Magazine* included seven tomato recipes. Her *Lady's Receipt-book,* published four years later, included four tomato recipes, and subsequent editions added another five. Miss Leslie's *New Cookery Book,* her culminating work, published in 1857, contained over thirty recipes with tomatoes as ingredients. In all, she published more than fifty different tomato recipes during her lifetime.[9]

While the quantity was notable, the quality and diversity of Leslie's tomato recipes were also significant. Tomatoes were stuffed, hashed, pickled, baked, scolloped, pickled, broiled, and preserved. They were added to several soups and gumbos and to many types of meats and poultry products including veal, ham, pork, beef, calf's feet, and sweetbreads. Tomato ketchup was added to many other dishes as well. While Leslie borrowed many of her recipes from other sources, including Mary Randolph's work, she drew on her vast culinary background to revise all of her tomato recipes. Many were altered from one edition of a cookbook to another, indicating her continuing quest to improve them according to her experience.

THE GROWTH OF TOMATO COOKERY

Tomato cookery was not limited to the eastern seaboard. In Ohio, Edward James Hooper's *Practical Farmer, Gardener and Housewife* contained twelve tomato recipes, most of which were taken from Maria Eliza Rundell,

Mary Randolph, Lydia Maria Child, N. K. M. Lee, Eliza Leslie, and the *American Farmer*. Hooper's two recipes for tomato soup merit attention. The first called for a broth base and eight or ten tomatoes and was seasoned with thyme, savory, marjoram, parsley, sliced onions, okra, salt, and pepper. Hooper suggested adding a spoonful of flour a few minutes before serving to thicken it. His "Tomato Mutton Soup" contained a rack of mutton, tomatoes, chopped onions, potatoes, turnips, and carrots and was seasoned with thyme and sweet basil. Hooper later became editor of the *Western Farmer and Gardener,* in which several additional tomato recipes were published.[10]

Lettice Bryan's *Kentucky Housewife* featured more than twenty tomato recipes, including some for baked, broiled, stewed, fried, and pickled tomatoes. Her recipes for baked tomatoes, tomato soy, and tomato marmalade were derived from Mary Randolph. Several recipes were similar to those published in other cookbooks, but others were fascinating deviations. A ketchup recipe called for horseradish and red wine. Her "Tomato Jumbles"—made of green tomato pulp—was used for seasoning and gravy. Her tomato soup recipes were closer to today's tomato soup than those previously published. Bryan recommended eating tomatoes raw for breakfast but considered them "also a fine accompaniment to roast meats, for a dinner."[11]

By the 1840s tomatoes were an important part of most cookery books. *Modern Cookery,* written by the British cookbook author Eliza Acton and edited by Sarah Hale for publication in the United States in 1845, contained several well-thought-out tomato recipes. One of the simpler ones was for "Tomatas en Salade," in which she recommended dressing sliced tomatoes with salt, pepper, oil, and vinegar. Despite its French-sounding name, Acton claimed that this was the American fashion of preparing tomatoes and that they were often served in England in just such a manner. Her "Tomata dumplings, or puddings" recipe, lifted from the *American Farmer,* alluded to the delicate spicy flavor of the tomato. In a postscript, Acton noted that

> It is possible that the tomata, which is, we know, abundantly grown and served in a great variety of forms in America, may there, either from a difference of climate, or from some advantages of culture, be produced in greater perfection than with us, and possess really "the delicate spicy flavour" attributed to it in our receipt, but which we cannot say we have ever yet discovered here; nor have we put its excellence for puddings to the proof, though some of our readers may like to do so.[12]

Acton's comment was the first British recognition of American tomato cookery, which, in a very short period of time, had surpassed British culinary practices.

In 1847 Sarah Rutledge's *The Carolina Housewife* gave directions for stewed, baked, fried, pickled, and preserved tomatoes. She also instructed her readers in making tomato sauce, salad, and jelly and using these in macaroni, fish, shrimp, and meat dishes. She had two tomato recipes for omelets, two that included okra, two for paste, two for ketchup, and three for preserved tomatoes. As one would expect, many of her recipes were similar to those previously published, but four were particularly intriguing. Her "Minced Meat" recipe began with chopped cold meat or fowl, which was fried in butter with seasonings and peeled tomatoes. Her "Meat and Potato Balls" consisted of chopped cold meat mixed with mashed potatoes, to which concoction cream, seasonings, and tomato ketchup were added. This mixture was then pressed into balls and served with a sauce. For her "Knuckle of Veal with Tomatoes," the cook was to stew beef and tomatoes over a slow fire for several hours. Her "Macaroni a la Napolitana" was among the first recipes published in America that combined macaroni with the tomato.[13]

Cookbooks were not the only means of purveying tomato cookery. Frequenting the pages of agricultural works and newspapers were recipes for tomatoes combined with almost every available food. The *American Farmer* published its first tomato recipe in 1822. During the next forty years, hundreds of techniques and recipes for cooking, eating, and preserving tomatoes, including Spanish, French, Turkish, Polish, Italian, and British methods, were published. In 1843 a single issue of the *Boston Cultivator* published twenty-five recipes for preserving and preparing tomatoes.[14]

Recipes published in agricultural journals and newspapers were rarely original. Editors usually lifted them from cookbooks, but there were several exceptions. Recipes for tomato ketchup, tomato dumplings, and tomato figs first appeared in agricultural sources and later were incorporated into cookbooks. Despite their general lack of originality, recipes published in newspapers and other periodicals promoted tomato cookery to a wide audience. Cookbooks largely targeted the middle and upper urban classes. As the vast majority of Americans lived and worked on farms during the early nineteenth century, the agricultural press and newspapers disseminated tomato cookery to a rural audience and boosted its legitimacy throughout America.

Magazines published tomato recipes beginning in the 1830s. *Lady's Book,* published by Louis Antoine Godey, began publishing tomato recipes in 1831. In 1837 Godey combined his *Lady's Book* with the *American Ladies' Magazine.* With Sarah Hale as its editor, the new journal, *Godey's Lady's Book,* was one of the most influential magazines in the mid-nineteenth century. Other periodicals, such as *Scientific American,* also published tomato recipes. *Scientific American* later identified them as "formulae," presumably in keeping with its scientific image.[15]

Tomato cookery was also loosely associated with many social reform efforts under way in America. Tomato recipes frequently appeared in temperance and vegetarian cookbooks, tracts, and newspapers. C. A. Neal's *Temperance Cook Book* included several tomato recipes, as did James Fisher's *Temperance House-keeper's Almanac* and Ann H. Allen's *The Housekeeper's Assistant,* which espoused temperance principles. Likewise, religious groups and other communities readily adopted the tomato. Tomato recipes appeared in Quaker cookbooks. Elizabeth Lea's *Domestic Cookery, Useful Receipts,* first published in Baltimore in 1845, incorporated tomatoes in nine recipes; subsequent editions included almost twenty recipes that made use of tomatoes. Shaker communities sold tomato seeds and also published recipes for tomato cookery. The *Shaker Gardener's Manual,* published in New Lebanon, New York, in 1843, recommended stewing tomatoes for pies, pickles, sauce, and ketchup. Vegetarian communities such as the Oneida Community in upstate New York not only adopted the tomato but also went into business canning, preserving, and selling tomato products. Likewise, socialist communities like the North American Phalanx in Monmouth, New Jersey, grew and canned tomatoes during the 1840s and 1850s.[16]

The Spanish, Italian, German, and French Influences

American tomato cookery did not develop in a vacuum. Many countries and cultures had previously absorbed the tomato into their culinary practices, and Americans regularly borrowed tomato recipes from others. Most cookbook authors borrowed from foreign or ethnic sources. As Karen Hess has noted, Mary Randolph probably acquired her Spanish recipes from her sister, Harriet Randolph Hackley, who had lived in Cádiz, Spain. Other recipes reputed to be Spanish in origin were regularly published in periodicals and cookbooks. Eliza Leslie published several Spanish recipes with tomatoes, such as "Pollo Valenciano." In *Novisimo Arte de Cocina,* the first Spanish-language cookbook published in the United States, tomatoes were integrated into many recipes. This cookbook was printed on a stereotype press for a client in Mexico and was probably not distributed in the United States.[17]

Not surprisingly, many Italian recipes published in America contained tomatoes. N. K. M. Lee included an Italian recipe for tomato sauce in *The Cook's Own Book.* Sarah Rutledge identified two of her tomato recipes as Italian, including one called "Macaroni a la Napolitana." In 1849, after a cruise in the Mediterranean, Captain Engle of the U.S. Navy sent home Italian recipes for preparing and preserving tomatoes. Heirom Aime, U.S.

Vice Counsel in Spezia, Italy, believed that larger "profits could be realized by [the tomato's] sale in our markets and from our stores, than from any other winter vegetable now to be obtained."[18]

German-language periodicals and cookbooks were published in the United States during the 1830s and 1840s, and many contained tomato recipes. *Ceres,* a German-language periodical printed in Lebanon, Pennsylvania, included a tomato recipe in 1839. George Girardey, who was of Swiss or Alsatian extraction, published the first German-language cookbook in America. His *Höchst nützliches Handbuch über Kochkunst,* published in Cincinnati in 1842, included five tomato recipes. This was a partial translation of his English-language *Manual of Domestic Economy,* published in 1841. Girardey was trained as a cook in Philadelphia and was a proponent of traditional Pennsylvania Dutch cookery. Another German-language cookbook, *Die Geschickte Hausfrau,* was published in Harrisburg, Pennsylvania, and had three recipes with tomatoes, including one for "Tomätoes Catsup."[19]

After the British, the most significant influence upon American cookery was French. This influence commenced in the late eighteenth century and grew steadily during the early nineteenth century. Charles Elmé Francatelli, Queen Victoria's maître d'hôtel, was born in London of Italian heritage and studied under the legendary French chef Marie-Antoine Carême. Francatelli's *French Cookery,* the first American edition of which was published in Philadelphia in 1846, included recipes for tomato sauce and tomato purée that were used in several other recipes. Francatelli later expanded his cookbook and added English, German, and Italian recipes.[20]

In the same year that Francatelli's cookbook was published, an adapted translation of Louis Eustache Audot's *La Cuisinière de la compagne et de la ville* was printed under the title *French Domestic Cookery.* Its tomato recipes included the usual ones for forcing and preserving tomatoes and for their use in soups, sauces, and gazpacho. In addition, Audot added tomato sauce to herring, beef fillet, veal, cauliflower, larded eel, mutton, sheep's tongues, brains, tails, lamb, lamb's heads, pork, pigs' ears, fish, macaroni, and other dishes.[21] The French influence was also channeled through American cookbooks, which often contained recipes identified as French in origin, such as Ude's tomato sauce recipe. Many more French recipes were included in cookbooks, periodicals, and newspapers without attribution.

RESTAURANTS

The French influence was particularly manifested in America's restaurants. Most Americans ate at home during the late eighteenth and early nineteenth century. Taverns, saloons, public houses, and inns served mainly travelers. Restaurants were opened toward the end of the eighteenth century, mostly by French refugees. Their clientele consisted of businessmen and an increasingly affluent upper class. Tomatoes were served in these restaurants at least by the 1820s and probably much earlier.

In 1827 Joseph Collet opened a French restaurant on the lower floor of his Commercial Hotel on Broad Street in New York. In a footnote to an advertisement in the *Post* on behalf of his establishment, he promulgated his preserved tomatoes. Eight years later Collet sold his hotel and restaurant to John, Peter, and Lorenzo Delmonico. The Delmonico brothers had established a cafe and pastry shop in New York in 1827, and ten years later they opened what became known as Delmonico's Restaurant. From its inception it offered a variety of tomato dishes.[22]

Large hotels with restaurants developed in most large American cities, such as Washington, D.C., Philadelphia, Baltimore, and Boston. A new era of luxury was launched with the opening of the Tremont Hotel in Boston, completed in 1829, and the Astor House in New York, which opened in 1836. At first, hotels offered set meals for everyone. Food was served on long tables at specific hours. During the 1820s they discarded the American plan with its fixed meals and offered à la carte menus, or European plan service. Menus announced the food to be served each day. Some hotels won fame as gastronomic centers.[23] Most hotel restaurants purportedly had French chefs. As one contemporary observer pointed out, however, many "French" chefs were born in New York, Pennsylvania, or Massachusetts.

Tomatoes were noted on hotel menus as early as 1825 and were probably served prior to this date. In 1830 the Exchange Coffee House served "Fricandeaux aux Tomata." By the 1840s the diversity of tomato dishes dramatically increased. The New York's Bank Coffee House sold tomatoes for 6 1/2 cents. The Astor House served tomatoes stuffed, baked, raw, and stewed. The Atlantic Hotel's bill of fare added "Tomate Sauce" to larded chicken and veal cutlets. The Troy House advertised mutton chops with tomato sauce. The City Hotel of New-York's Table D'Hôte served tomato sauce with "Fricandeau," "Riz-de-Veau," "Côtelettes de Veau,"mutton chops, and sweetbreads.[24]

Tomato cookery was not limited to New York restaurants. In Washington, D.C., Birch's U.S. Hotel served tomato soup and roast duck with tomatoes. Cincinnati's Gibson House served "Sweetbreads, with a New Tomato Sauce," "Young Quails, Larded, with Tomato Sauce," fresh stewed tomatoes, and pickled tomatoes. San Francisco's Ontario Hotel served tomatoes as a vegetable. In New Orleans tomatoes were served at the St. Charles Hotel. On the Mississippi River they were served on steamboats. In

Missouri the St. Louis Hotel served "Grenadin de Veau, with sauce tomate," stuffed tomatoes, and stewed tomatoes. In Boston the Pavilion served tomato sauce on "Cotelette de Mouton." Boston's Revere House served tomatoes from its inception in 1847. At the dinner celebrating its grand opening, tomato soup and larded sweetbreads with tomato sauce were served.[25]

At Barnum's Restaurant in Charleston, Virginia, tomatoes were baked so that they were browned all over without breaking the skins. A correspondent for the *Charleston Evening News* had "seen a small man eat a peck or more at a sitting—in round quantities—then turn in upon the raw." When he ordered tomatoes, the waiter corrected his English: "Tomato! Oh tomatusses, you mean." "Yes," said the correspondent, "it must be tomatusses!" Thereupon, with an indulgent smile, the waiter supplied him with the baked tomatoes.[26]

From the 1820s on, tomatoes were considered a delicacy and were served on special occasions for visiting dignitaries. When New York's Governor De Witt Clinton visited Chillicothe, Ohio, in 1825, tomatoes were included in the fare. Charles Dickens was served tomato sauce with "Filet de Boeuf" and "Fricandeau" at the City Hotel when he visited New York in 1842. When President James K. Polk visited St. Louis in 1849, he was served "Grenadin de Veau" with "sauce tomate," stuffed tomatoes, and stewed tomatoes. When the Imperial Majesty the Sultan of Turkey Amin Bey visited Boston in 1850, the merchants of the city fed him "Cotelette de Pigeons" with tomato sauce.[27]

WAYS OF PREPARING TOMATOES

Pop historians and culinary experts alike have frequently proclaimed that prior to the Civil War tomatoes were cooked for three hours to counter their purported poisonous qualities. In 1845 the *Indiana Farmer* maintained that, to use tomatoes in omelets, one should cook them for more than three hours before adding the eggs. Two years later Sarah Rutledge borrowed the recipe and picked up the refrain, although she also included recipes in her cookbook for raw tomatoes and for tomatoes cooked for much shorter periods. Eliza Leslie too claimed that tomatoes needed to be cooked for a long time, "otherwise they will have a raw taste, that to most persons is unpleasant." Along with this advice, she included recipes for raw tomatoes. In 1860 *Godey's Lady's Book* reported that tomatoes were delicious and wholesome but were often spoiled by the manner in which they were served: "It is not one time in a hundred more than half cooked; it is simply scalded, and served as a sour porridge. It should be cooked three hours—it cannot be cooked in one." These suggestions were made for reasons of taste; they had nothing to do with the purported poisonous qualities of the tomato. Despite its statement, *Godey's Lady's Book* also published other tomato recipes containing raw tomatoes or tomatoes cooked for shorter durations.[28]

Throughout the early nineteenth century, accounts of eating raw tomatoes regularly appeared in a variety of American sources. Some raw tomato aficionados recommended seasoning them with sugar, molasses, vinegar, salt, pepper, mustard, or milk. The most common way to eat raw tomatoes was sliced and seasoned, like cucumbers, with vinegar, salt, and pepper. Others plucked them from the vine and ate them "like ripe fruit, without seasoning." The *Cultivator* recommended just eating them "as fast as you can."[29]

The normal methods of cooking available at the time were used in tomato cookery: frying, boiling, baking, roasting, and broiling. During the early nineteenth century, all of these methods were accomplished over an open fireplace as few Americans owned cast-iron stoves. Cooking over the open fire was "a back-breaking, dangerous, and inefficient task." Regulating the heat, maneuvering heavy pots and kettles, and controlling the fire and smoke were only a few of the chores that were necessary for this type of cooking. Stoves were generally adopted during the 1830s, but cooking by the fireplace continued in many parts of the country until after the Civil War. Compared with cooking on a gas stove, the wood- or coal-burning stove was "a crude and cumbersome thing, but contrasted with the open fire at floor level it was a major release and delight for the housewife."[30]

The general adoption of the cast-iron stove made it possible for the middle-class housewife to expend greater time and energy on sophisticated recipes rather than on the tough and grueling mechanics of cooking. Likewise, the general adoption of the icebox improved the housewife's ability to preserve food. It was no accident that tomato cookery in America grew simultaneously with the adoption of the stove and the icebox. It was also no accident that cookbooks often contained advertisements for stoves and iceboxes.

The earliest European mode of cooking tomatoes, published in 1544, was to fry them in olive oil and season them with salt and pepper. Variations of this recipe were published regularly in Europe and later in the Americas. Mary Randolph fried sauce-like tomatoes with shredded meat or fowl. The *New England Farmer* recommended slicing up green tomatoes and frying them in butter. Lettice Bryan coated them with bread crumbs before frying them.[31]

Stewed tomatoes were served alone or combined with other products. In England, Philip Miller and others noted that Italians, Spaniards, and Jews stewed tomatoes for use in sauces and soups. Maria Eliza Rundell recommended stewing tomatoes for ten or twelve minutes along with vinegar, salt, and pepper. By the early nineteenth century, stewed tomatoes were "very much admired" in America, and recipes for them were regularly published beginning in 1829. In Catherine Esther Beecher's recipe, they were placed on toast. Eliza Leslie recommended stewing them with beef and lamb. They were served in restaurants throughout the 1840s and 1850s.[32]

Baking, or cooking with heat in an enclosed space, could be done in the fireplace or in an oven. In the fireplace, a dish was set down and covered with a kettle. Coals or embers were then placed on top. Oven baking was sometimes, but not always, done outside the house to minimize the danger of fire. Coals or embers were placed in a chamber to heat it up and then removed before the food was placed inside. Tomatoes were baked in several different ways. Maria Eliza Rundell simply sliced them in half, seasoned them, and baked them in an oven. A second method was reflected in Sarah Rutledge's recipe in which tomatoes and bread crumbs were placed in a dish and baked together. This process was similar to scolloping tomatoes, which was popularized by Mary Randolph. She baked them in layers with other products, such as bread crumbs, butter, and seasonings. Baked vegetables, particularly tomatoes, became a traditional dish on the southern table and were found on the menus of many restaurants.[33]

Originally marmalade was a Portuguese quince jam with chunks of fruit. Culinary historian C. Anne Wilson reports that it was infused into English cookery in the sixteenth century. Oranges soon replaced quinces as the major ingredient in the English marmalade, although other products were also used. Randolph published two recipes for tomato marmalade. Her "Tomata Marmalade" called for the pulp of green tomatoes seasoned with pepper, salt, cloves, and garlic. Her "Sweet Marmalade" was the same except that the pulp was mixed with loaf sugar and then stewed until it was the consistency of jelly. The first recipe was used for seasoning gravies. The uses of the second were not stated; however, the ingredients in it were a little closer to those in other marmalade recipes. Versions of these recipes were published throughout the nineteenth century in newspapers, cookbooks, and periodicals, including *Godey's Lady's Book,* and continued into the twentieth century, as a few tomato marmalade recipes published in culinary magazines and cookery books attest. Unlike Randolph's tomato marmalade recipes, these were intended for use as preserves.[34]

Tomatoes were made into preserves, jellies, and jams. A correspondent in Pennsylvania's *Northumberland Public Aspect* noticed a new kind of tomato preserve prepared by the landlady at Mr. Pardoe's inn and thought its flavor "remarkably rich and fine." Lettice Bryan's recipe for tomato jelly was simple: boil tomatoes with the rind of a lemon, drip the mash through a thin jelly bag, and boil it with loaf sugar and lemon juice. In 1840 the *American Farmer* offered an even simpler recipe: strain stewed tomatoes through a coarse cloth, add an equal part of sugar, and boil for a few hours. Sarah Rutledge's recipe for tomato jelly required brown sugar and egg whites. Her recipe for tomato preserves called for combining green tomatoes with oranges or lemons but without any sugar. In 1843 the *Boston Cultivator* offered a process for making tomato jam: rub stewed tomatoes through a sieve, add the same weight in sugar, "and stew away to the usual consistence of jam."[35]

Eggs and tomatoes were a favorite combination in the Caribbean and Europe by the mid-eighteenth century. By 1796 tomato omelets were noted in private letters in Louisiana, where they had probably been served previously for years. According to Mary Randolph's recipe, the first one published in America that combined eggs and tomatoes, the process was to fry chopped onions in butter with seasoning, put in the tomatoes, and, when they were nearly done, mix in the eggs. Tomato omelet recipes were published frequently and were very popular.[36]

Tomatoes were combined with many other vegetables. One of the most common was okra, which originated in West Africa. It was observed in Brazil in 1658, in Surinam in 1686, and in Philadelphia in 1748. Thomas Jefferson listed okra as growing in Virginia along with tomatoes in 1781. It may have been brought to America by slaves from the Caribbean or directly from Africa. Some recipes for gumbo, an okra-based Creole soup or stew, incorporated tomatoes as an ingredient. Whether the combination of tomatoes and okra originated in the Caribbean, in New Orleans, or on a plantation in the American South is unknown. In any case, recipes for gumbo, okra soup, and other such dishes were published regularly throughout the nineteenth and twentieth centuries.[37]

Tomatoes were also used to make alcoholic beverages. A correspondent in the *Southern Agriculturist* asserted that the leaves of the tomato plant could be used for making beer. As these were toxic, it is not surprising that this was the only report of beer's being made from the tomato plant, although later in the nineteenth century southerners made beer from the tomato fruit. When tomato whisky made its debut, the editor of the Albany *Plow* propounded that it was the latest product "being tortured out of this innocent vegetable" and facetiously predicted that tomato whisky would "hereafter be quoted among the agricultural products of the country."

Tomato champagne made a brief appearance in the 1850s. The editor of the *East Tennessean,* who considered himself a good judge, pronounced it "a first rate article."[38] Despite such rave reviews, tomato champagne was promptly forgotten.

Recipes for tomato wine were published beginning in the 1840s. These were basically tomato juice with sugar added to promote fermentation without the addition of yeast. Other ingredients were occasionally added. One recipe noted that tomato wine was "very much improved by adding a small proportion of the juice of the common grape." This wine was purported to be "peculiarly adapted to some diseases and states of the system" and was "particularly recommended for derangements of the liver." Wellington Rose from the Hancock Shaker Village in Pittsfield, Massachusetts, manufactured tomato wine for the "sick and infirm." According to W. Bacon from Richmond, Massachusetts, the wine so closely resembled old Madeira that "it would have troubled an amateur to detect the difference." Mrs. E. F. Haskell affirmed in 1861 that tomato wine was "much thought of in some places." Not everyone, however, was enchanted with it. A correspondent for the *Working Farmer* professed that he failed to discover why tomato wine was entitled "to such an appellation," as he believed it was only a "fruitful source for the consumption and waste of sugar." A correspondent in the *American Agriculturist* stated, "We have seen a miserable, sweet, alcoholic liquid made from tomatoes, and we thought it an insult to a respectable vegetable to put it to such use." Despite these critics, tomato wine recipes are still with us, as occasional recipes in food magazines and recent tomato cookbooks demonstrate.[39]

RECIPE INGREDIENTS

Almost all commonly available spices, herbs, and condiments were used in tomato recipes: horseradish, parsley, cloves, nutmeg, garlic, mustard, ginger, allspice, and many others. A great diversity of vegetables was combined with tomatoes, such as okra, onions, carrots, potatoes, and cabbage. Dairy products and grain products were also ingredients in tomato recipes, as were meats, eggs, poultry, fish, and other seafood products of all available types.

As can be seen from the earliest tomato recipes, sugar was often added to cut down the tomato's acidic taste. Early-nineteenth-century recipes called for sugar sparingly, but as the decades passed, the amount dramatically increased. In part this increase was due to the decline in the price of sugar imported from the West Indies. It flooded into the American market and its consumption escalated spectacularly. This was clearly reflected in tomato recipes for pies, tarts, jams, marmalades, preserves, and many other products. By 1834 sugar was used in equal weight to that of the tomatoes. Mrs. Major Gano offered a recipe for preserving tomatoes that required equal quantities of ripe tomatoes and white sugar. Her method was to alternate layers of sugar and tomatoes in a jar until it was nearly full. Likewise, Lettice Bryan's first recipe for tomato preserves called for equal weights of tomatoes and brown sugar. Her second recipe was for preserving yellow tomatoes by using an equal weight of loaf sugar. C. A. Neal preserved green tomatoes by boiling them in a solution of sugar and water. The assortment of tomato recipes using sugar also enlarged as the century progressed. *Cookery, on a Simple and Healthful Plan* recommended stewed tomatoes with sugar. A correspondent to the *American Agriculturist* claimed that tomatoes required more sugar to make them palatable. Some cooks mixed sugar with dried tomatoes to create a medicine. Tomato sauce recipes started including sugar beginning in 1828. The first known ketchup recipe with sugar was published in 1844.[40] By the 1840s sugar was an ingredient in almost half of the published tomato recipes.

Alcohol was another important ingredient in tomato cookery. It was added to tomato recipes for a variety of reasons. Brandy and wine were frequently added to tomato sauce and ketchup recipes for preserving purposes. In a revised edition of Maria Eliza Rundell's *A New System of Domestic Cookery,* Emma Roberts added a glass of port wine for coloring purposes to her recipe for using tomatoes in making an "imitation of Guavas." *The Lady's Annual Register* added red wine to stuffed tomato for reasons of taste, as did Mme Utrecht-Friedel's *The French Cook.*[41]

Preserving and Canning Tomatoes

Like most fruits and vegetables, fresh tomatoes were limited to their season, which varied from a month or two in the far North to several months in the deep South. Whatever the length of the season, however, at its height tomatoes were available in profusion. The solution was to preserve them when they were abundant for use in the off-season. Many were the ways of preserving tomatoes, including making ketchup as well as drying, bottling, and canning them.

Ketchup recipes were first published in England during the early eighteenth century. British explorers, colonists, and traders came into contact with a ketchup forerunner, a fermented soy-based fish sauce, while they were in Asia. Upon their return to Europe, they attempted to duplicate it.[42] The concoction takes its name from the term *kētsiap,* from the Amoy dialect of Chinese. By the seventeenth century, different types of *kētsiap* were used throughout East and Southeast Asia. Its major claim to fame was its purported longevity, great enough, according to some recipes, to survive the time it took for a ship to sail from Britain to India. Other cookbook authors claimed that their recipe would keep for two, three, seven, or even twenty years. Many ketchup recipes called for additives that contributed to preservation, such as vinegar, salt, and alcoholic beverages. Some employed preserving techniques like covering the contents of the bottle with melted butter or keeping it cold. These recipes produced condiments that were sour in taste, mainly because of the addition of vinegar, wine, or mum (beer). Their consistency was thin because the pulp was usually strained out. As soybeans were not grown in Europe, British cooks used such substitutes as anchovies, mushrooms, walnuts, and oysters.

Tomato ketchup may have originated in America. It was widely used throughout the United States in the late eighteenth and early nineteenth century. Francis Vigo made a type of ketchup in the Old Northwest. In the early nineteenth century, Mrs. E. P. Gaines made it in Alabama. In 1804 Dr. James Mease divulged that tomato ketchup was made by French immigrants in Philadelphia. Eight years later Mease published the first known recipe for "Tomatoes or Love-Apple Catchup." It contained few of the hallmarks of traditional ketchup—for instance, no vinegar or anchovies. The product was not strained and was therefore thick. As tomato ketchup became the rage, new ketchup-like recipes were developed. Mrs. Cadwallader, a member of Philadelphia's elite, used "soy tomata ketchup" in a recipe for stewing fish. She gave the recipe to Mrs. Coxe before 1818. Culinary historian William Woys Weaver believed that Mrs. Cadwallader received this recipe from the Binghams, who had a French cook as early as the 1780s.[43]

Unlike American authors, British cookbook writers maintained a more traditional approach to making tomato ketchup. The 1817 edition of Dr. William Kitchiner's *Apicius Redivivus; or, the Cook's Oracle* included a recipe for "Tomata Catsup" with anchovies. The pulp was strained out. The following year he published another recipe, which lacked anchovies but included malt vinegar. Unlike Mease's recipe, this was passed through a sieve twice, producing a thin, juice-like extract.[44]

In America, tomato ketchup caught on fast. A second American recipe appeared on the final page of the 1818 edition of John Gardiner and David Hepburn's *The American Gardener.* As both Gardiner and Hepburn were dead before this edition was published, the author of the recipe is unknown. It was a strained version of previously published tomato sauce recipes, but its author proclaimed that tomato ketchup was better than mushroom ketchup for culinary purposes. Similar judgments were rendered by others. In 1827 the *American Farmer* published a recipe that it considered superior to any West

Indies ketchup. It included salt, peppercorns, allspice, mustard, and eight pods of red pepper. The concoction was simmered in vinegar for three or four hours and then strained through a wire sieve and bottled. While it was usable in two weeks, it improved "much by age." This was recommended as a remedy for dyspepsia (indigestion). Chili pepper had been an ingredient in tomato recipes from the earliest accounts in pre-Columbian cookery. Later, the combination of chili pepper and tomato ketchup was renamed chili sauce. Other combinations of chili and tomatoes were called salsa, one of the fastest-growing product lines in America today.[45]

During the early nineteenth century, the terms *tomato sauce* and *tomato ketchup* were used almost interchangeably, a phenomenon that raises the issue of the distinction between them. Ketchup was intended to survive long periods of time and therefore possessed more ingredients thought to be preservatives, such as vinegar. Additional preserving techniques were incorporated—for example, filling the top of the jar with some substance thought to be a preservative or tying a bladder around it. Many tomato sauce recipes, although not all, were served shortly after their preparation. Tomato sauce recipes frequently included ingredients that easily spoiled, such as dairy products. Many tomato ketchup recipes, although not all, were strained once or twice and were therefore very thin. When the pulp was strained out, the resulting product was yellow in color, not red. These principles of distinction, however, were not always adhered to in practice. In any case, both tomato sauce and ketchup were immensely popular. When U.S. Senator Robert Henry Goldsborough died in 1836, the inventory of his estate noted eighteen bottles of tomato ketchup laid up in the larder. Hundreds of tomato ketchup and sauce recipes were published throughout the nineteenth century.[46]

Pickling was another form of preserving tomatoes. Tomatoes had been pickled since at least the early nineteenth century. During the 1830s Dr. John Cook Bennett pickled green tomatoes, but he preferred the ripe fruit, which he thought was "highly medicinal, and has a much better flavor." Following *The Lady's Book*'s recipe for "Ripe Tomato Pickles," one pricked them with a fork, seasoned them, and packed them in a jar filled with vinegar and water, layered between sliced onions and spices. Eliza Leslie suggested substituting a larger quantity of spice for the onions. A correspondent in the *Cultivator* recommended using mustard seed and horseradish. A correspondent in the *Michigan Farmer* suggested using cinnamon and cloves. The *Prairie Farmer* recommended pouring over them "sugar dissolved in vinegar, in the ratio of a pound of sugar to a pint of good vinegar." Almost everyone recommended soaking them an hour or so in fresh water just before they were to be used. Likewise, tomatoes were chopped up and mixed with other ingredients and were pickled in vinegar. These tomato relishes were sometimes called hodge-podges. Tomato pickles and pickle relish appeared on hotel menus by the 1840s.[47]

Another way of preserving tomatoes was through drying. Dr. James Dekay, a physician living on Long Island, visited Turkey and noted that the Turks dried tomatoes in the sun until they became a thick paste. Perhaps because of its exotic origin, this recipe was reprinted frequently for more than ten years. In Maryland the *Cambridge Chronicle* disclosed that this method preserved "the true flavor of the fruit for several years." As only the essence was left after evaporation, only a small quantity was needed to flavor a dish. A correspondent to the *Cultivator,* using Dekay's recipe, reported that "a specimen of what was prepared for my family more than twenty months ago, is still in a fine state of preservation." He also recommended that it be used "to impart a relish to a dish of *Pilaf,* that is, rice cooked with meat or butter, or soup, for a large family." According to a recipe in the *Boston Cultivator,* dried tomatoes should be packed away in bags, "which should then be hung up in a dry room." When they were wanted for use, they were soaked in warm water. The *Southern Agriculturist* offered three sensible recommendations: first, boil them; second, if they are dried in the sun, make sure that they are protected from insects and weather; and third, use the oven for drying them if necessary.[48]

The aforementioned Captain Engle of the U.S. Navy, who sent home Italian recipes for preparing and preserving tomatoes, explained that Genoa grocery stores sold tomato conserve as thick as lard and "sufficiently stiff to stand in conical forms, as loaf sugar." It was brought in from the country by the wagon load. He saw barrels from a five-horse wagon unloaded and exported to Russia and America. In Spezia, the conserve "resembled a large Bologna sausage and was much darker than the conserve prepared at Genoa."[49]

New ways to preserve tomatoes for winter and spring use were still needed. The beginning of the commercial preserving industry is usually traced to Nicholas Appert's bottling experiments in France. One of these involved stewing, straining, and sealing tomatoes in bottles. The bottles were then boiled in a water bath. Appert included no salt, sugar, spices, or other preservatives. He had "no doubt that this new method will give valuable and economical results." One immediate economical result was that a grateful French government awarded him 12,000 francs for his discoveries. His book explaining these techniques was translated for publication in Great Britain in 1810 and published in the United States two years later.[50]

Tomatoes were bottled commercially in the United States during the 1820s and 1830s. In New York the French-born Joseph Collet put up tomatoes "in a manner that he guaranteed would keep even in the warmest climates for twelve months." He considered the bottled tomatoes important especially for the sea traveler who needed bottled food for long voyages. In 1830 an advertisement in the *New England Farmer* promoted the sale of "Tomato Mustard" and "Tomato Ketchup" in glass bottles for fifty cents or thirty-three cents, depending on size. Bottled tomato ketchup was marketed in New York by 1834 and in Hartford four years later.[51]

The most prominent early bottler was William Underwood, who brought knowledge of Appert's bottling processes from Great Britain in 1817. He established a factory in Boston in 1822 but did not bottle tomatoes until 1835, when he sold them in two-pound bottles at a cost of three and a quarter dollars. Each bottle of his "Hermetically Sealed Tomatoes" contained the substance of about two dozen tomatoes and kept "any length of time." It was prepared by "straining the seeds and skins from the tomatoes and evaporating the watery particles by slow heat." By 1839 his tomato ketchup, bottled in Boston, was shipped to New Orleans, Mobile, and Pensacola. From New Orleans it was shipped throughout the Mississippi River system.[52]

In 1847 Harrison W. Crosby, from Jamesburg, New Jersey, filled tin pails with tomatoes, boiled the pails, and sealed the top of each with a tin disk. He packed six pails to a box and sold them in the Washington Market in New York City. To gain visibility for his procedures, Crosby sent samples to Queen Victoria, President Polk, U.S. senators, newspaper editors, restaurants, and hotels. Two years later the *New York Tribune* declared of the canned tomatoes that "whatever the secret of their preparation, we are bound to acknowledge that their preservation has not impaired their flavor. They taste as they would have tasted when plucked from the vines."[53]

By the late 1840s other factories in New Jersey canned tomatoes. John Bucklin established a factory at the North American Phalanx in Red Bank. Bucklin saved some canned tomatoes, opened and ate them forty years later, and reported that they tasted just as good as they did the day they were canned. In Newark, Harry Evans established a factory for Kensett and Company, where tomatoes were packed for Dr. Elisha Kane's Arctic Expedition in 1851. By the 1850s factories that canned tomatoes could be found in New York City, Astoria on Long Island, and the Oneida Community in upstate New York.[54]

The *Bangor Whig and Courier* encouraged farmers to can their own tomatoes, as did Georgia's *Southern Cultivator,* the *Maine Farmer,* and Missouri's *Valley Farmer.* The procedures for canning tomatoes were laborious. The fruit was peeled and cut up, salt and pepper were added, and the mixture was put into two-quart tin canisters, each having a hole two inches in diameter in the top. A circular piece was soldered over the opening, leaving an anvil hole. The canisters were placed in a kettle of boiling water for twenty minutes. The anvil hole was then stopped up with a spile of pine. When the canisters had cooled, the spile was cut even with the tin, and melted sealing-wax was placed over it.[55] E. F. Haskell's *Housekeeper's Encyclopedia,* published in 1861, offered extensive advise on canning "Hermetically-sealed Tomatoes." Haskell advised cooks to check the cans the day after they were preserved. If they were not drawn in, they were imperfect, or the hot air was not fully expelled. In either case, the fault was corrected by repeating the sealing process.[56]

Canning was difficult under any conditions, but particularly so in urban areas. The need for easier means of preserving food in the home led to the creation of a variety of new techniques and devices. In 1855 Robert Arthur took out a patent for a grooved-ring can. He created Arthur, Burnham & Company in Philadelphia and began producing "patent, air-tight, self-sealing cans and jars," which became known as "Arthur's cans." They came in several sizes and were made of a variety of substances, including tin, glass, and stoneware. They had an open top with a lid that extended over the top part of the jar and was sealed with cement prepacked in its lip. According to *Godey's Lady's Book,* these were easier to use than other available jars and were used by thousands of families, hotels, and boarding houses. Arthur, Burhnam & Company offered extensive instruction for canning tomatoes. They recounted that during the summer season the tomato was "found upon the table every day" and was given up "with regret when it is down by the frost." Their solution was, of course, to use Arthur's cans and jars.[57]

From the consumer's standpoint, the disadvantage of Arthur's cans was that they were not easily reusable. After the contents were removed, it was difficult to reglue the jars. Although Arthur did claim that other substances could be packed in the groove, such as cloth, leather, or even newspapers, one wonders how effective the seal was. His jars were soon overshadowed by the invention of New Jersey-born John L. Mason, who had set up a metalworking shop on Canal Street in New York. On November 30, 1858, when he was twenty-six years old, Mason patented the self-sealing zinc lid and glass jar. It revolutionized the preservation of fruits and vegetables in the home. As it was easy to use and comparatively inexpensive to produce, it became very popular. By 1860 the mason jar was shipped throughout the United States. One of the first cookbooks to incorporate it was Haskell's *Housekeeper's Encyclopedia.*[58]

By the late 1850s and early 1860s, the commercial canning of tomatoes had become a significant business. Edmund Morris reported that one canner in southern New Jersey employed thirty people (mainly women) and produced 50,000 tomato cans in a single season. The canner purchased tomatoes when the price dropped to twenty-five cents a bushel. At the high season, a hundred and fifty bushels arrived at the cannery every day, which required everyone to work far into the evening to keep up. Morris described the operation:

> The building in which the business is carried on was constructed expressly for it. At one end of the room in which the canning is done is a range of brick-work supporting three large boilers; and adjoining is another large boiler, in which the scalding is done. The tomatoes are first thrown into this scalder, and after remaining there a sufficient time, are thrown upon a long table, on each side of which are ten or twelve young women, who rapidly divest them of their leathery hides. The peeled tomatoes are then thrown into the boilers, where they remain until they are raised to a boiling heat, when they are rapidly poured into the cans, and these are carried to the tinmen, who, with a dexterity truly marvelous, place the caps upon them, and solder them down, when they are piled up to cool, after which they are labelled, and are ready for market. The rapidity and the system with which all this is done is most remarkable, one of the tinmen soldering nearly a hundred cans in an hour.[59]

Along with commercialization came adulteration. Some of it was unintentional. As there was no lining in the cans, acidic contents inevitably ate into their inner wall. Other adulterations were intentional. In 1860 tomato grower James Gregory of Marblehead, Massachusetts, described an interaction with a Boston dealer who had made a ton of tomato ketchup using only dried apples:

> What a humbug this honest looking man is, was my instantaneous thought. But Chemistry here interposed, and said, "Not so fast, not so fast! What's in a name? Is not the characteristic acid of the tomato and the apple the same? viz: malic acid; and in dried or preserved fruits is not the flavor dependent almost wholly on the characteristic acid, most of the more delicate flavors of the fresh fruit being too subtle to be

> retained by such processes? Now the quantity of malic acid in ripe tomato exceeds that in the ripe apple; but when the apple is dried, and particularly when it becomes very dry by age, and the acid thus greatly concentrated, may not the proportionate difference be lessened, and thus in all essential characteristics your apple ketchup will become tomato ketchup?"[60]

In 1866 cookbook author Pierre Blot warned against what was sold under the name of ketchup, believing that "many cases of debility and consumption" came from "eating such stuff." Examinations of some commercial ketchup showed that it contained excessive preservatives, including boric acid, formalin, salicylic acid, and benzoic acid. Also, coal tar was used to create a bright red coloring thought necessary to catch the eye of the consuming public. Other ketchup contained harmful bacteria and was "filthy, decomposed and putrid." These unsanitary conditions could be explained. The tomato season was relatively brief—from mid-August until mid-October. Ketchup manufacturers could not make their entire year's supply of tomato ketchup during this short period, so many of them concentrated tomato pulp during the season and preserved it for future use. Investigations showed that the conditions for storage were unsanitary and that the pulp was handled carelessly, as indicated by the presence in abundance of molds, yeasts, spores, and dead bacteria. However, these problems were unnecessary and avoidable. The Edward C. Hazard Company of Shrewsbury, New Jersey, proudly produced ketchup without preservatives and without contamination, as did the H. J. Heinz Company in Pittsburgh, Pennsylvania.[61]

For good reason, suspicion of canned products continued well into the twentieth century. While the canning process improved dramatically after the Civil War, abuses were not dealt with effectively until the passage of the Pure Food and Drug Act in 1906. Despite these difficulties and lingering suspicions, bottled and canned preservation techniques increased the availability of tomato throughout the year. By the last decades of the nineteenth century, all but the poorest Americans could easily afford canned tomato products.

NOTES

1. Hannah Glasse, *The Art of Cookery Made Plain and Easy* (London: Printed for the Author, 1758), 341; Richard Briggs, *The New Art of Cookery* (Philadelphia: W. Spotswood, R. Campbell, and B. Johnson, 1792), 80; Alexander Hunter, *Culina Famulatrix Medicinae; or, Receipts on Modern Cookery*

(York: T. Wilson and R. Spence, 1804), 128–30, 184, 193.

2. Maria Eliza Rundell, *A New System of Domestic Cookery* (New York: R. M'Dermut & D. D. Arden, 1814), 127; A. F. M. Willich [Thomas Cooper, ed.], *The Domestic Encyclopedia: or a Dictionary of Facts and Useful Knowledge* (Philadelphia: Published by Abraham Small, 1821), vol. 3, 114; William Kitchiner, *The Cook's Oracle* (Boston: Munroe and Francis, 1822), 250.

3. Manuscript cookbook, Mrs. George Read, New Castle, 1813, Holcomb Collection, Historical Society of Delaware, Wilmington; *American Farmer,* 5 (September 26, 1823): 215.

4. Louis Eustache Ude, *The French Cook* (London: J. Ebbers, 1813), 37.

5. Richard Alsop, *The Universal Receipt Book or Complete Family Direction by a Society of Gentlemen in New York* (New York: I. Riley, 1814), 45.

6. Mary Randolph, *The Virginia House-wife* (Washington: Davis and Force, 1824), 31, 33, 34–35, 56–57, 95–97, 107, 201; (Washington: Way & Gideon, 1825), 92–93, 140–41, 230.

7. Lydia Maria Child, *The Frugal Housewife* (Boston, March & Capen and Carter & Hendee, 1829), 31–32, 114–15; Richard J. Hooker, *The Book of Chowder* (Boston: Harvard Common Press, 1978), 9, 104.

8. Child, *Frugal Housewife,* 31–32; N. K. M. Lee, *The Cook's Own Book* (Boston: Munroe & Francis, 1832), 222–23.

9. Mrs. Eliza Leslie, *Seventy-Five Receipts for Pastry, Cakes, and Sweetmeats* (Boston: Munroe and Francis, 1834), 103–4; (Boston: Munroe & Francis, 1851), 107–8, 118; Leslie, *Domestic French Cookery, chiefly Translated from Sulpice Barué* (Philadelphia: Carey & Hart, 1832), 21–22, 32–33, 35, 37–38, 73–74; Leslie, *Directions for Cookery* (Philadelphia: E. L. Carey & A. Hart, 1837) 15–18, 32–33, 83–84, 95–96, 103–4, 110–12, 177, 200, 223–25; *Miss Leslie's Magazine,* 2 (September 1843): 102–3; (October 1843): 138; (December 1843): 211; Leslie, *The Lady's Receipt-book* (Philadelphia: Carey and Hart, 1847), 44–45, 66–67, 90–91; Leslie, *Miss Leslie's Lady's New Receipt-book* (Philadelphia: A. Hart, Late Carey & Hart, 1850), 372–75, 382, 386–87, 405, 433, 445, 445–46; Leslie, *Miss Leslie's New Cookery Book* (Philadelphia: T. B. Peterson, 1857), 22–23, 39–41, 64–66, 75–76, 152–54, 156–59, 176, 178, 190–91, 211–12, 255, 280, 292–94, 323, 326, 365, 369–70, 374–75, 397–400, 547, 558–60, 581.

10. Edward James Hooper, *The Practical Farmer, Gardener and Housewife* (Cincinnati, Ohio: Geo. Conclin, 1839), 495.

11. Lettice Bryan, *The Kentucky Housewife* (Cincinnati: Stereotyped by Shepard & Stearns, 1841), 24–25, 168, 171–72, 174, 183, 216–18, 367, 426.

12. *American Farmer,* 3d ser., 4 (August 17, 1842): 101; Eliza Acton, *Modern Cookery in All its Branches . . . Prepared for American Housekeepers. By*

Mrs. S. J. Hale (Philadelphia: Lea and Blanchard, 1845), 108, 117–18, 241–42, 281.

13. Sarah Rutledge, *The Carolina Housewife* (Charleston: W. R. Babcock, 1847), 111–12.

14. *American Farmer,* 4 (September 1822): 208; *Boston Cultivator,* 5 (September 9, 1843): 282.

15. *Scientific American,* 3 (September 9, 1848): 408.

16. Miss C. A. Neal, *Temperance Cook Book* (Philadelphia: Temperance and Tract Depository, 1835), 28, 36; Ann H. Allen, *The Housekeeper's Assistant Composed upon Temperance Principles* (Boston: James Munroe, 1845), 73, 74, 78-79; James Fisher, *Temperance House-keeper's Almanac* (Boston: James S. Fisher, 1843); Elizabeth E. Lea, *Domestic Cookery, Useful Receipts* (Baltimore: H. Colburn, 1845), 30–31, 88, 101; (Baltimore: Cushings and Bailey, 1853), 47–89, 134, 142, 145–46, 160, 163–65; *A Shaker Gardener's Manual* (New Lebanon, N.Y.: United Society, 1843; reprinted by Applewood Books, Cambridge, Massachusetts), 21–22; *Oneida Circular,* 3 (April 1, 1854); 4 (August 23, 1855): 124; North American Phalanx, label for canned goods, dated by hand, 1851.

17. Karen Hess, "Historical Notes and Commentaries," in Mary Randolph, *The Virginia House-wife* (Columbia: University of South Carolina Press, 1983), xxxii; *American Farmer,* 14 (November 2, 1832): 270; 4th ser., 1 (September 1845): 88; *Western Farmer and Gardener,* 5 (January 1845): 15; Rutledge, *Carolina Housewife,* 103; Leslie, *Miss Leslie's Lady's New Receipt-book,* 440–41; *Novisimo Arte de Cocina* (Philadelphia: Estereotipado é impreso por Compania, 1845); William Woys Weaver, "Additions and Corrections to Lowenstein's Bibliography of American Cookery Books, 1742–1860," *Proceedings of the American Antiquarian Society,* 92 (1982): 363–77.

18. Lee, *Cook's Own Book,* 222–23; Rutledge, *Carolina Housewife,* 111–12; letter from Heirom Aime, U.S. Vice Consel, Spezia, dated March 15, 1849, as in the *Germantown Telegraph,* September 4, 1850.

19. *Ceres,* 1 (September 1839): 32; George Girardey, *Höchst nützliches Handbuch über Kochkunst* (Cincinnati: Stereotyped by F. U. James, 1842), 5, 6, 112; *Die Geschickte Hausfrau* (Harrisburg, Pa.: Lutz und Scheffer, 185?), 11–12, 27–28; letter from William Woys Weaver to the author, dated June 3, 1993.

20. Richard Osborn Cummings, *The American and His Food: A History of Food Habits in the United States* (Chicago: University of Chicago Press, 1940), 31; Ude, *French Cook,* 40; William Kitchiner, *The Cook's Oracle* (New York: J. & J. Harper, 1830), 239; Leslie, *Domestic French Cookery;* Charles

Elmé Francatelli, *French Cookery. The Modern Cook* (Philadelphia: Lea and Blanchard, 1846), 20, 40–41, 372.

21. Louis Eustache Audot, *French Domestic Cookery* (New York: Harper & Brothers, 1846), 57, 64, 98–102, 105–6, 144, 164, 186, 257, 262, 264–65, 268, 272, 308.

22. *New-York Evening Post,* December 14, 1827; Delmonico's Restaurant, 1837, Carte du Restaurant Français, des Freres Delmonico, 2–4, New-York Historical Society.

23. Frederick Marrayat, *A Diary in America,* part 2 (Paris: Baudry's European Library, 1840), 40; Richard J. Hooker, *Food and Drink in America, A History* (Indianapolis & New York: Bobbs-Merrill, 1981), 141–42.

24. Madeira Hotel, July 25, 1825, Bill of Fare of a Dinner Given by the Citizens of Chillicothe, and its Vicinity in Honor of the Guest of Ohio, Gov. De Witt Clinton of N. York, New-York Historical Society; Exchange Coffee House, September 17, 1830, Menu, New York Public Library; Bank Coffee House, 1840, Bill of Fare, New-York Historical Society; Astor House, October 11, 1849, Bill of Fare, New-York Historical Society; Atlantic Hotel, December 31, 1843, Bill of Fare, New-York Historical Society; Troy House, February 22, 1848, Washington Birth-Day, Bill of Fare, New-York Historical Society; City Hotel of New-York, August 22-5, 1844, Table D'Hôte, New-York Historical Society.

25. Birch's U.S. Hotel, July 11, 1846, Gentleman's Ordinary, New-York Historical Society; Gibson House, February 14, 1849, Table D'Hote, New-York Historical Society; Ontario Hotel, 1849, Bill of Fare, New-York Historical Society; St. Charles Hotel, December 22, 1846, Fifth Anniversary Dinner of the New England Society of Louisiana, as in the *Daily Picayune,* December 23, 1846; Magnolia [Riverboat], April 9, 1846, Bill of Fare, as in Norbury Wayman, *Life on the River* (New York: Crown, 1971), 228; St. Louis Hotel, March 22, 1849, Public Dinner given to Honorable James K. Polk, New-York Historical Society; Pavilion, August 28, 1844, Bill of Fare, New-York Historical Society; Revere House, May 19, 1847, Dinner to Celebrate Opening of the Revere House, New-York Historical Society.

26. *Charleston Evening News,* as in the *New England Farmer,* 4 (February 1852): 76.

27. Madeira Hotel, July 25, 1825, Bill of Fare of a Dinner Given by the Citizens of Chillicothe, and its Vicinity in Honor of the Guest of Ohio, Gov. De Witt Clinton of N. York, New-York Historical Society; City Hotel of New York, February 18, 1842, Bill of Fare in Honor of Charles Dickens, Esq., New-York Historical Society; St. Louis Hotel, March 22, 1849, Public

Dinner given to Honorable James K. Polk, New-York Historical Society; Revere House, November 4, 1850, Dinner to Amin Bey in Honor of his Imperial Majesty the Sultan of Turkey by the Merchants of Boston, New-York Historical Society.

28. *Indiana Farmer and Gardener,* 1 (June 28, 1845): 178; Rutledge, *Carolina Housewife,* 103–4; Leslie, *Lady's Receipt-book,* 44–45; *Godey's Lady's Book,* 51 (August 1855): 174; 57 (August 1858): 76; 61 (July 1860): 74.

29. Eliza Acton, *Modern Cookery in All its Branches . . . Prepared for American Housekeepers. By Mrs. S. J. Hale* (Philadelphia: Lea and Blanchard, 1845), 241; *Boston Cultivator,* 5 (September 9, 1843): 282; *Farmers' Cabinet,* 8 (September 15, 1843): 44–45; Bryan, *Kentucky Housewife,* 217; *Cleveland Herald,* August 2, 1835; *Cultivator,* 6 (September 1839): 134; Astor House, October 11, 1849, Bill of Fare, New-York Historical Society.

30. Hooker, *Food and Drink;* 95–96.

31. Randolph, *Virginia House-wife* (Davis & Force), 97, 107; *New England Farmer,* 14 (October 14, 1835): 106; Bryan, *Kentucky Housewife,* 217.

32. Philip Miller, *The Gardeners Dictionary* (London: Printed for the Author, 1748); *A Supplement to Mr. Chambers's Cyclopedia: or, Universal Dictionary of Arts and Sciences* (London: Printed for W. Innys and J. Richardson, 1753), vol. 1; Rundell, *A New System of Domestic Cookery* (London: John Murray, 1824), 259; Bernard M'Mahon, *The American Gardener's Calendar* (Philadelphia: Printed by B. Graves for the Author, 1806), 319; Child, *Frugal Housewife,* 31–32; Bryan, *Kentucky Housewife,* 216; Acton, *Modern Cookery,* 241; Allen, *Housekeeper's Assistant,* 73; Catherine Esther Beecher, *Miss Beecher's Domestic Receipt Book* (New York: Harper and Brothers, 1846), 78–79; Leslie, *Domestic French Cookery,* 32–33; Leslie, *Directions for Cookery,* 111–12, 200; St. Louis Hotel, March 22, 1849, Public Dinner Given to Honorable James K. Polk, New-York Historical Society.

33. Rundell, *American Domestic Cookery* (Baltimore: Fielding Lucas, Jun. R. J. Matchett, 1819), 140; Rutledge, *Carolina Housewife,* 103; Randolph, *Virginia House-wife* (Way & Gideon), 140; City Hotel of New-York, August 22-5, 1844, Table D'Hôte, New-York Historical Society.

34. C. Anne Wilson, *The Book of Marmalade* (New York: St. Martins/Marek, 1985), 123; Wilson, "Marmalade in New Found Lands," *Petis Propos Culinaires,* 18 (1984): 34–39; Randolph, *Virginia House-wife* (Way & Gideon), 201–2; *Godey's Lady's Book,* 59 (October 1859): 368.

35. *Northumberland Public Aspect,* as in the *Genesee Farmer,* 3 (September 7, 1833): 287; Bryan, *Kentucky Housewife,* 367; *American Farmer,* 3d ser., 2 (October 7, 1840): 155; *Boston Cultivator,* 5 (September 9, 1843): 282; Sarah Rutledge, *Carolina Housewife,* 164, 168.

36. Edward Long, *The History of Jamaica,* new edition with a new introduction by George Metcalf (London: Frank Cass, 1970), 773; John Lunan, *Hortus Jamaicensis* (Jamaica: Office of the St. Jago de li Vega Gazette, 1814), 234; Randolph, *Virginia House-wife* (Davis and Force), 107; *American Agriculturist,* 5 (September 1846): 269; *Letter Diary of Joseph Delfau de Pontalba to Wife,* September 17, 1796, Louisiana State Museum Historical Center, New Orleans.

37. Ulysis P. Hedrick, ed., *Sturtevant's Edible Plants of the World* (New York: Dover, 1972), 302; Randolph, *Virginia House-wife* (Davis and Force), 34–35, 95–96.

38. *Southern Agriculturist,* 5 (July 1832): 390; M. E. Porter, *Mrs. Porter's New Southern Cookery Book* (New York: Arno Press, New York Times Company, 1973), 183; Jonathan Periam, *The Home and Farm Manual* (New York: Greenwich House, 1984), 889; *Plow,* 1 (June 1852): 193; *East Tennessean,* as in the *Michigan Farmer,* 15 (November 1851): 351.

39. *Prairie Farmer,* 5 (July 1845): 168; *Ohio Cultivator,* 1 (September 1845): 133; *Horticulturist,* 8 (December 1853): 431; *Working Farmer,* 4 (January 1, 1853): 244; Mrs. E. F. Haskell, *The Housekeeper's Encyclopedia* (New York: D. Appleton, 1861), 277; *American Agriculturist,* 30 (August 1871): 286.

40. *New England Farmer,* 7 (December 19, 1828): 174; *Cincinnati Mirror,* 3 (September 6, 1834): 375; *American Farmer,* 2d ser., 1 (November 11, 1834): 223; *Farmer and Gardener,* 2d ser., 1 (November 11, 1834): 223; Bryan, *Kentucky Housewife,* 366–67; *New England Farmer,* 7 (December 19, 1828): 174; *American Agriculturist,* 1 (June 1842): 90–91; *Cookery, on a Simple and Healthful Plan* (Boston: George W. Light, 1842), 27; *Southern Planter,* 4 (September 1844): 204.

41. Rundell, *A New System of Domestic Cookery*, ed. Emma Roberts (Philadelphia: Carey and Hart, 1844), 192; *Fisher's Improved House-keeper's Almanac* (Philadelphia: R. Magee, 1850); *The Lady's Annual Register, and Housewife's Almanac, for 1843* (Boston: T. H. Carter, 1843), 59; Mme Utrecht-Friedel, *The French Cook* (New York: Wm. H. Graham, 1846), 59–60.

42. Terrien de Lacouperie, "Ketchup, Catchup, Catsup," *Babylonian and Oriental Record,* 3 (November 1889): 284–85; Lacouperie, "The Etymology of Ketchup," *Babylonian and Oriental Record,* 4 (February 1890): 71–72; Eneas Sweetlands Dallas, *Kettner's Book of the Table: A Manual of Cookery Practical, Theoretical, Historical* (London: Dulau, 1877), 266.

43. Mrs. Samuel Whitehorne is credited with an earlier tomato catsup recipe that appears in a handwritten manuscript titled *The Sugar House Book* with the date 1801 on the cover. The actual date of the recipes is much later.

James Mease, ed., *A. F. M. Willich's The Domestic Encyclopedia* (Philadelphia: William Young Birch and Abraham Small, 1804), vol. 3, 506; *Country Gentleman,* 19 (May 15, 1862): 318; Mrs. Gaines Diary, as in the *Mobile Daily Register,* July 6, 1879; Mease, ed., *Archives of Useful Knowledge* (Philadelphia: David Hogan, 1812), vol. 2, 306; recipe in the Coxe Family Papers in the Historical Society of Pennsylvania, as in a letter from William Woys Weaver to the author, dated April 28, 1990.

44. William Kitchiner, *Apicius Redivivus; or, the Cook's Oracle* (London: Samuel Bagster, 1817); *Apicius Redivivus* (London: Samuel Bagster, 1818), 480–81.

45. John Gardiner and David Hepburn, *The American Gardener* (Washington, D.C.: Joseph Milligan, 1818), 348; *American Farmer,* 9 (August 31, 1827): 191.

46. Maryland State Archives, *J. P. Liber #19, Inventories and Accounts of Sales 1836–1838.*

47. Gardiner and Hepburn, *American Gardener* (Washington, D.C.: Samuel H. Smith for the Authors, 1804), 27; *Cleveland Herald,* August 22, 1835; *Godey's Lady's Book,* 2 (March 1831): 168; Eliza Leslie, *Directions for Cookery* (Philadelphia: E. L. Carey & A. Hart, 1837), 223–24; *Cultivator,* 6 (September 1839): 134; *Michigan Farmer,* 1 (July 15, 1843): 87; *Prairie Farmer,* 3 (September 1843): 210; *Southern Agriculturist,* 10 (April 1837): 190-2; Allen, *Housekeepers' Assistant,* 79; Gibson House, Cincinnati, February 14, 1849, Table D'Hote, New-York Historical Society.

48. *Southern Agriculturist,* 4 (February 1831): 81–82; *Cambridge Chronicle,* September 26, 1835; *Cultivator,* 6 (November 1839): 183; *Boston Cultivator,* 5 (September 9, 1843): 282.

49. Letter from Heirom Aime, U.S. Vice Consel, Spezia, dated March 15, 1849, as published in the *Germantown Telegraph,* September 4, 1850.

50. Nicholas Appert, *The Art of Preserving* (New York: D. Longworth, 1812), 52.

51. *New-York Evening Post,* December 14, 1827; *New England Farmer,* 8 (February 12, 1830): 240; *Northern Courier,* February 1, 1838; December 6, 1838; Bunker [New York], May 13, 1834, New-York Historical Society.

52. *The Second Century* (Watertown, Mass.: William Underwood, 1927); canning label, as in Mary B. Sim, *Commercial Canning in New Jersey: History and Early Development* (Trenton: New Jersey Agricultural Society, 1951), 14; *Arkansas State Gazette,* November 13, 1839; *Pensacola Gazette,* April 14, 1838; *Quincy Whig,* May 18, 1839; *Illinoisan,* July 20, 1839; *Sangamo Journal,* November 3, 1838.

53. *Merchants' Review,* 24 (August 14, 1891): 7–8; *American Grocer,* 48 (July 20, 1892): 8; *New York Tribune,* as reported in David Bishop Skillman, *The Biography of a College* (Easton, Penn.: Lafayette College, 1932), vol. 1, 179–80.

54. Label at the Monmouth County Historical Association in the North American Phalanx Archive; *Merchants' Review,* 24 (August 14, 1891): 7–8; Sim, *Commercial Canning,* 16; *American Agriculturist,* 15 (July 1856): 233; Lewis Gannett, *Cream Hill; Discoveries of a Weekend Countryman* (New York: Viking, 1949), 68; *Oneida Circular,* as in Constance Noyes Robertson, *Oneida Community Profiles* (Syracuse. N.Y.: Syracuse University Press, 1977), 52.

55. *Bangor Whig and Courier,* September 21, 1849; *Maine Farmer,* 17 (October 11, 1849); *Southern Cultivator,* 11 (August 1853): 242; *Maine Farmer,* 23 (August 9, 1855); *Valley Farmer,* 7 (September 1855): 400; *Bangor Whig and Courier,* September 21, 1849; *Maine Farmer,* 17 (October 11, 1849).

56. Haskell, *Housekeeper's Encyclopedia,* 131.

57. U.S. Patent No. 12,153; *Godey's Lady's Book,* 55 (June 1857): 566; *Fresh Fruits and Vegetables All Year at Summer Prices; and How to Obtain Them* (Philadelphia: Arthur, Burnham & Gilroy, 1857), 32–34.

58. Julian H. Toulouse, *Fruit Jars: A Collectors Manual* (Camden, N.J.: Thomas Nielson & Sons, and Everybodys Press, 1969), 340–47; Haskell, *Housekeeper's Encyclopedia,* 131.

59. Edmund Morris, *Ten Acres Enough* (New York: J. Miller, 1864), 120–61.

60. *New England Farmer,* 12 (March 1860): 146.

61. Pierre Blot, *What to Eat, and How to Cook It* (New York: D. Appleton, 1866), 25; A. W. Bitting, *Appertizing or the Art of Canning* (San Francisco: Trade Pressroom, 1937), 670.

6

Tomato Medicine

In November 1834 Dr. John Cook Bennett declared that tomatoes could be used successfully in the treatment of diarrhea, violent bilious attacks, and dyspepsia. They were also good, he stated, for citizens traveling west or south, as they would save them "from the danger attendant upon those violent bilious attacks to which almost all unacclimated persons are liable." Bennett urged all citizens to eat tomatoes as they were "the most healthy article of the Materia Alimentary." He recommended that tomatoes supplant calomel (mercurous chloride) because their effects were less harmful. Calomel, a toxic substance, was used frequently by physicians as a purgative and for other purposes. Bennett predicted that "a chemical extract" would soon be obtained from tomatoes that would "altogether supersede the use of Calomel in the cure of diseases." He also asserted that "the free use of the Tomato" made a person "much less liable to an attack of Cholera, and that it would in the majority of cases prevent it."[1]

Bennett was born in Massachusetts in 1804, but he lived most of his early life in Ohio. He was licensed to practice medicine after a three-year preceptorship with another physician and the successful completion of a six-hour examination. Later he attended lectures at the medical department of McGill College in Canada. While in Montreal, he visited the "Scotch and French Royal Colleges of Physicians and Surgeons" and their hospitals, where he found tomatoes highly recommended "as a collateral agent" and used extensively "by convalescent persons and valetudinarians." His interest in tomatoes was piqued by an article in the *Cincinnati Farmer and Mechanic* in July 1834 that covered a variety of topics related to the cultivation and use of the tomato. The editor opined that the tomato's purported injurious effects could be attributed to using it in a "green state." The "wholesomeness" of the fruit, however, he deemed to be a subject "properly belonging to the medical faculty."[2]

A few weeks after the article appeared, Bennett was elected president of the faculty of the medical department of Willoughby University of Lake

Erie, located in Chagrin, about twenty miles east of Cleveland. At the time, Dr. William Smith, a physician from Monroe County, Michigan, and a professor at Willoughby University, was experimenting with the tomato as a medicine.[3] Bennett summarized Smith's beliefs and presented them to his class in his opening lecture.

To promote the university and himself, Bennett promptly published his lecture in the *Painesville Telegraph* under the pseudonym of MEDICUS. He received only one response, from an pseudonymous IATROS. IATROS, which in Greek means "physician," was an individual very knowledgeable about botany and medicine. He was fully convinced of the wholesomeness of the tomato. He wondered, however, "in what manner are the alleged deobstruent effects of the tomato produced?" He then commenced an extensive discussion of botany and ended with some comment on quacks who roamed about "pouring drugs of which they know little into bodies of which they know less." Busily launching Willoughby University, Bennett did not respond, and nothing further was heard from him regarding tomatoes for several months. For a variety of reasons he was fired from his position at Willoughby University in April 1835.[4]

A few months later Bennett began circulating extracts of his introductory lecture on the virtues of tomatoes to newspapers and agricultural journals. His claims were not unusual, and similar claims had been made for many other vegetable plants including lobelia, sarsaparilla, mustard, dandelion, and rhubarb.[5] However, there were Americans who still considered the tomato to be inedible. Others who knew they could be eaten chose not to eat them. Even those who ate tomatoes were unaware of the purported healthful qualities claimed by Bennett. That the tomato was the most healthful vegetable was a surprise to most Americans, and it was therefore newsworthy.

On August 1, 1835, the *Ohio Farmer and Western Horticulturist* reprinted Bennett's claims. Within a week they were published in Pennsylvania, New York, and Virginia. During the following weeks the article appeared in newspapers in Alabama, New Jersey, Ohio, North Carolina, South Carolina, Florida, and Massachusetts. On August 21 Bennett wrote to the *Daily Cleveland Herald,* gleefully reporting that his findings about the tomato were "going the general round of publication, and have had the effect to awaken the public mind to an investigation of the merits of this invaluable exotic." His letter also contained additional information about the healthful qualities of the tomato. For instance, he professed that it was "an invaluable prophylactic or preventive, against Asiatic or Asphyxiated Cholera." He also included a testimonial by Dr. Robley Dunglison, who had served as Thomas Jefferson's personal physician. Bennett added fuel to the interest by publishing recipes for raw tomatoes, tomato sauce, fried tomatoes, ketchup, and pickles. In his tomato pickles recipe he recommended using "the green fruit, by the same process that you would observe in the pickling of cucumbers, or other articles. The ripe fruit may likewise be pickled; and in fact, it is the preferable article; as it is in that case highly medicinal, and has a much better flavor." Bennett also pointed out that the tomato was a "fashionable dessert." These recipes were published when the tomato season was at its height and therefore had maximum impact. Bennett forwarded copies of this second article to additional newspapers and journals. The *Maine Farmer* published it in late August, the *American Farmer* on September 1, and the *New England Farmer* on the following day. The *Genesee Farmer,* the *Tennessee Farmer,* the *Cultivator, Southern Agriculturist,* and *Farmers' Register* published his lecture and recipes in the following weeks. Many other newspapers and agricultural, medical, and religious journals promptly picked up Bennett's claims.[6]

The immediate response was electric. The editor of the *Thomsonian Recorder* congratulated Bennett on his success in finding that tomatoes might replace calomel "after a search of four thousand years, under the guidance of 'scientific principles,' in the 'discovery' that at least one poison may *possibly* yet be supplanted by an article whose effects on the system are in harmony with life." The *New York Transcript* facetiously reported that Bennett's claims had been received,

> like most improbable things, with mouths wide open, and the most delectable credulity, by the public. The papers have copied it, their readers have believed it, and the people have been all agog to cure their liver complaints, their diarrhoea, and their dyspepsia, by the use of the tomato. Some had them prepared in one way and some in another. But many devoured them raw.[7]

While Bennett's claims were published in all regions of the United States, they received much more coverage in the northern states. The most likely reason for this difference was that southerners were already familiar with tomatoes and did not believe that they possessed magical qualities, though many southerners already considered them healthy.

In late September 1835 Bennett moved to Erie, Pennsylvania, and attempted to launch the Sylvanian Medical School. While in Erie he wrote to several prominent Americans, sending newspaper articles about his findings. On November 30, 1835, he published a request for more information about the tomato and more recipes. He promised that he would cite the source of all information in an extended article on the tomato to be published in the spring of 1836 and would send a copy of the finished article to each person responding to his request.[8]

Bennett was particularly curious as to the chemical contents of the tomato. Lieutenant George Morell of the United States Engineers reported that "several unsuccessful attempts had been made to analyze the tomato, although its principal constituents are the same as other vegetables." Thomas G. Clemson from Lafayette College in Pennsylvania had commenced an examination of the tomato, but when he found that it would require much more time and attention than he could devote, he "relinquished the undertaking." Clemson was under the impression that malic acid appeared to be always present in the fruit. Constantine Rafinesque, a self-proclaimed doctor who had nostrums for many illnesses, had previously recommended tomatoes as a remedy for consumption. He responded to Bennett's letter by reporting that, besides "tomatic acid," the fruit contained "mucilage, water, and an extractive and coloring matter." He claimed that the tomato was "everywhere deemed a very healthy vegetable, and an invaluable article of food." David Thomas, editor of the *Genesee Farmer,* avowed that tomatoes were very helpful to persons recovering from fever. Edward Hall Barton, dean and professor of the Medical College of Louisiana, read Bennett's remarks about calomel with great interest and stated in response that "if subsequent experience shall sustain your position of its having a specific influence on the liver and being a substitute for calomel, you will confer lasting benefit upon your country, and erect an enduring monument to your reputation."[9]

Bennett's attempt to launch the Sylvanian Medical School failed, and he resigned because of ill health and moved to Hocking City, Ohio, where he continued to write about tomatoes. He published the testimonials he had received in the *Genesee Farmer* in 1836. These were in turn picked up by other periodicals. A year later Bennett finished a series of articles on the history, culture, and hygiene of tomatoes and published it first in the *Hocking Valley Gazette* and later in the *Botanico-Medical Recorder* in Columbus. He concluded after a thorough study of ancient texts that there was "no portion" of the world where the tomato was not indigenous in some of its varieties. He opposed training tomatoes upon poles as they were never intended "by God or Nature" to be trained up on anything and should therefore lie on the ground. Bennett quoted the British *Cyclopedia or Universal Dictionary* as saying that the tomato had the "reputation of being stimulant, or aphrodisiacal." This was the only serious reference to the tomato's reputed aphrodisiacal qualities published in America before the twentieth century, and it reflected outmoded beliefs in England, not America.[10]

In late 1837 Dr. A. J. Holcombe credited Bennett with introducing the tomato "as a medicine, in its crude state." Bennett boasted that he had been the first person in the United States to recognize the healthful aspects of the tomato, even though he knew it was not true. Many physicians, including the previously noted de Sequeyra, Samuel Green, Alexander Hunter, William Kitchiner, and James Mease, had promoted them earlier. The tomato's medicinal properties had been endorsed in Continental Europe since the sixteenth century. American physicians who had received their medical education in Europe were aware of its reported healthful qualities. Throughout the late eighteenth and early nineteenth century there were physicians who challenged the view that the tomato was unhealthy or inedible.

Beginning in the early 1830s the agricultural press systematically promoted the tomato's healthy image. For instance, in 1831 Horatio Spafford reported that ingesting tomato sauce removed headaches, a bad taste in the mouth, "straitness" of the chest, and painful heaviness in the liver and that it improved the action of the bowels. Bennett had published Spafford's claims for the healthful qualities of the tomato but had failed to cite the original date of publication, which superseded his own claims.[11] Bennett's information about the virtues of the tomato derived from William Smith.

Though he failed to credit others for their ideas, Bennett was an effective promoter and deserves credit for popularizing the edible tomato in America. Although it was not ascertained until the twentieth century, consuming one hundred milligrams of raw tomatoes provides twenty-three milligrams of vitamin C, or about forty percent of the adult recommended daily allowance (RDA), and about nine hundred international units of vitamin A, or about thirty percent of the adult RDA. While heating at ordinary cooking temperatures does not destroy vitamin A, heating and oxidation do destroy vitamin C. Eating raw tomatoes, as Bennett recommended, was therefore healthier than eating tomatoes cooked for long periods as recommended by many cookbook authors. Tomatoes also contain small amounts of potassium, calcium, iron, sodium, thiamine, and riboflavin. As Americans in general were vitamin deficient at the time, Bennett's tomato campaign could only have improved the health of those who took his advice.

Bennett moved to Illinois in 1838. Two years later he joined the Mormons, then residing in Nauvoo. He quickly became an intimate of Joseph Smith, the leader of the Mormon church, who happened to be a vegetarian. Bennett republished his articles about the tomato in the *Nauvoo Times and Seasons,* adding that "as health is essential to our happiness and prosperity as a people, we would earnestly recommend its culture to our fellow-citizens, and its general use for culinary purposes. Do not neglect it."

Later excommunicated by Joseph Smith, Bennett left the Mormon Church under a cloud of scandal. But the Mormons heeded his advice about the tomato, encouraging their members to make free use of it, as it possessed, "in an eminent degree, the virtues of calomel divested of the deleterious qualities, by which they will, in most cases, avoid all those harassing bilious affectations, and obstructions, to which unacclimated persons are so frequently subjected." Bennett wrote one additional article on the tomato and encouraged journals to reprint his previously published articles, but his campaign on behalf of the tomato ended in 1841.[12]

Bennett had predicted that "a chemical extract" would soon be obtained from tomatoes that would "altogether supersede the use of Calomel in the cure of diseases." Two months after this claim received wide attention, Dr. A. J. Holcombe began advertising tomato pills in Greensboro, Alabama. Three months later, he announced their availability through the *Botanico-Medical Recorder.*[13] His extract of tomato pills, he said, possessed "hepatic, cathartic and diuretic qualities." Not much is known about Holcombe's operation, but it paled by comparison to those of Archibald Miles from Ohio and Guy R. Phelps from Connecticut.

THE AMAZING ARCHIBALD MILES

Archibald Miles was born in New York in 1804 but moved to Cleveland with his family while he was still a youth. In 1824 he moved to Brunswick, Ohio, where he was employed as a merchant. He became an agent for British Hygiene, a restorative medicine developed in Great Britain. In 1836 he marketed American Hygiene Pills, a product of his own creation. During the spring of 1837 Miles met an unidentified physician, probably John Cook Bennett, who suggested that he change the name of his American Hygiene Pills to Extract of Tomato Pills. Miles moved from Brunswick to Cincinnati with his family, taking with him several barrels of his pills.[14] In Cincinnati he set up a "laboratory" and, within a few weeks of his arrival, advertised Dr. Miles' Compound Extract of Tomato.

According to their advertisements, Miles and his associates had spent years and fortunes developing the pills. They specifically claimed to have been working on tomato experiments since 1833 and to have expended $30,000 developing their scientific medicine. Miles declared that he had tested his medicine on various diseases, and the result of each well-substantiated case was duly published. Finally, he claimed, he had succeeded in "extracting a substance from the *tomato,* which from its peculiar effect upon the hepatic or biliary organs" he "denominated *Hepatine.*" Hepatine produced "all the beneficial results of Calomel without the possibility of producing the deleterious consequences common to that article." Hepatine's greatest powers, he went on, were particularly manifested "upon the organs of *secretion* and *excretion.*" Despite their dramatic efficacy, he assured his readers, the pills acted "in perfect harmony with the known *laws* of life" and were, "undoubtedly, one of the most valuable articles ever offered for public trial or inspection."[15]

These pills were attractively packaged. Adorning each box was a portrait of Dr. Miles, which, according to Miles, had cost $1,000 to engrave. Boxes came in two sizes: one with forty-five pills, priced at fifty cents, and the other with a hundred pills, priced at one dollar. The pills were three grains apiece in size and came in two colors to denote their intended use: the whites were cathartic, and the yellow ones were tonic.[16] According to the directions in the box, three white pills taken at bedtime produced a mild cathartic operation by morning: "Some persons require five or six, and others only two." One white pill, the directions claimed, "given once in 3 or 4 hours will produce an alterative effect." When a speedy operation was required, the patient needed merely to pulverize or dissolve the pills. Yellow pills were intended to restore vigor and tone to the system. Seven or eight of them were said to produce free perspiration and operate moderately as a cathartic. For weak or debilitated stomachs, the directions continued, "after giving a dose of the white pills, one yellow pill taken will restore tone and keep up a proper action."

In the first stages of a common cold, Miles explained, "equal quantities of the white and yellow pills should be taken together, sufficient to keep the bowels open until the complaint is removed." For anyone suffering from "bilious fever, inflammation of the head or chest, rheumatism, pleurisy, or any other form of acute disease," he suggested a "full dose of the white pills, say 6 or 10, and repeat every six or eight hours, until they operate freely." For chronic affections of the liver and dyspepsia, five or six of the white pills were recommended at bedtime, the same number of the yellow the next evening, and on the following evening four of the white again. To produce an alterative effect, "excite a healthy biliary secretion, and restore the digestive organs to the natural tone," two of the white pills should be "taken every evening, for three or four weeks; and every other day, one of the yellow pills morning, noon, and evening." During this course, the patient should abstain from eating "hot cakes, and all indigestible food." In cases of scrofula and syphilis, Miles recommended the same course as for liver complaints and dyspepsia. In fact, he pronounced, at "the first stages of disease in almost every form, it exceeds all former discoveries in medicine, either from the vegetable or mineral kingdoms." Even young infants could take the medicine, albeit at smaller doses.[17]

Miles swiftly established a network of agents in the midwestern and southern states. Some were hired at an annual salary and were located in "the most important parts of the country." Miles concentrated on locating individuals interested in becoming wholesalers who were responsible for a particular geographical area. In turn, wholesalers developed local retail outlets. Agents, wholesalers, and retailers advertised the tomato pills in newspapers and through circulars they distributed to customers. By late 1837 the pills were "advertised in almost every part of the country." By 1838 Miles's network of general agents and retailers extended from New Orleans, Mobile, and Houston along the Gulf Coast to Maine along the Canadian border; and from Charleston, Baltimore, and Boston along the Eastern Seaboard to as far west as Davenport, Iowa.[18]

Miles's promotion campaign was enhanced by the publication of articles in newspapers, periodicals, and journals across the country. In an 1837 article announcing that the pills were for sale, the *Western (Cincinnati) Christian Advocate* quoted extensively from advertisements and usage directions. This article was picked up and reprinted by the *Maine Farmer* and the *Southern Agriculturist.* F. W. Chester, the editor of the *Cincinnati Journal and Western Luminary,* took "some pains to inquire of medical men" who had used Miles's medicine and believed that his "extract of the Tomato will prove a substitute for Calomel in a great variety of cases." The physicians regarded it as "a great blessing to the human family." Bishop John Purcell, editor of Cincinnati's *Catholic Telegraph,* was inclined to think that Miles's tomato pills would "stand the test of scrutiny, and prove a most desirable acquisition to the world, and particularly to the people of this country." Dr. Gamaliel Bailey, editor of the *Philanthropist* and a practicing physician, had no hesitation in introducing the product to the notice of his readers, noting that Dr. Miles, "in conjunction with a few other physicians, has devoted himself to the investigation of the nature and the medical properties of the Tomato, from which he at length succeeded in procuring an extract, of excellent purgative and alterative properties." Bailey had seen "some favorable notices of its virtues in the papers of our contemporaries." He heard it "highly praised by those who have used it, as a grateful, searching and efficient cathartic."[19]

The *Troy (Ohio) Times* announced that, to its certain knowledge, more than six hundred physicians used the medicine. The editor perceived that "the prejudices which many scientific physicians have heretofore entertained against the properties and use of the Tomato medicine" were yielding to "a conviction in its great utility in the cases recommended by its proprietors."

In Kentucky the *Louisville Daily Herald* concluded that Miles' Extract could be recommended to the public with safety and propriety because it was "scientifically prepared" and its "component parts" were "freely made known to all medical men." It was "recommended by men of science and professional skill, as a discovery calculated to benefit mankind." The editors of the *Jeffersonville Courier* added that they had conversed with persons who used Miles's pills, "and they invariably pronounced it a most efficacious medicine."[20]

These endorsements usually coincided with major advertising campaigns by Miles in the newspapers that endorsed them. It is probable that Miles offered to purchase advertising space provided that the editor favorably reviewed his tomato pills. Whatever the reason for their publication, these editorials gave legitimacy and visibility to Miles's pills.

The success of the promotional campaign was immediate. An independent observer claimed that Miles had "hundreds of thousands" of customers in the western and southern states alone. The demand was so great during the winter of 1837 that Miles had to close his shop because he ran out of ingredients to make his pills. He apologized to customers, particularly in the southern states and the British West Indies, as he was unable to fill all the orders he received. Miles was slower in establishing agents in the eastern states. Thomas Bell, a wholesale druggist, was hired as his agent in Philadelphia and was responsible for licensing retailers in Pennsylvania, Delaware, and Maryland. Charles Hosmer, a merchant, was selected for Hartford, Connecticut. Miles's pills were shipped to him in October 1837, and Hosmer published advertisements for them shortly thereafter. In November, Miles selected Hoadley, Phelps and Co., a wholesale drug firm, as his agent in New York City.[21]

In December 1837 Dr. A. J. Holcombe became upset with Miles's claim that he was the originator of tomato extract. Holcombe pointed to his own previous advertisements, which predated Miles's announcements by almost two years. John Cook Bennett immediately agreed with Holcombe. Alva Curtis, the editor of the *Botanico-Medical Recorder,* asked for the recipes from Miles and Holcombe. Holcombe cheerfully and gratuitously bestowed his recipe "upon the community at large." It was basically condensed tomato juice with the consistency of stiff tar. Miles gave Curtis his recipe for making his extract, excluding the process for making hepatine, but refused to let him publish it. Curtis claimed that if the pills were "made according to that recipe, they could not be a 'substitute for calomel,' as they could never produce the mischief effected by that deadly drug."[22]

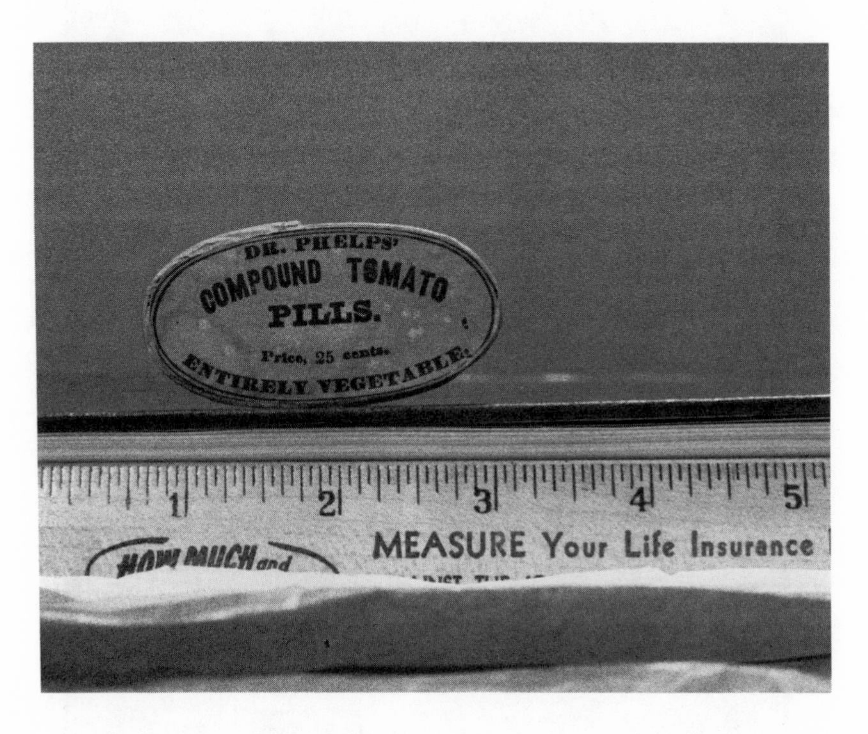

Dr. Phelps' Compound Tomato Pill Box. (Photo by the author.)

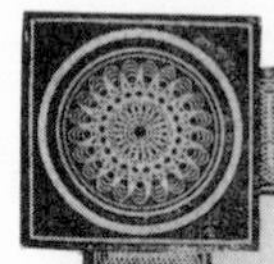

DR. MILES'

COMPOUND

EXTRACT OF TOMATO,

THE CELEBRATED SUBSTITUTE FOR CALOMEL.

After devoting several years, and the expenditure of a large sum of money in experiments, the proprietors have at length succeeded in extracting from the TOMATO (Solanum Lycopersicum) a vegetable principle producing all the beneficial effects indicated by CALOMEL, without any of its deleterious results. This principle, (Hepatine) is a mild and efficient Cathartic, acting at the same time as a Tonic, Diuretic, and Diaphoretic.

Its action upon the system is very general, no part escaping its influence; it is, however, upon the secretories and execretories that its great power is particularly manifested; from this it will be seen to have a direct effect upon the biliary organs, and to be particularly adapted to the treatment of bilious fevers and other diseases where a torpidity or congestion of the liver and portal circle prevail: hence the great success that has attended its administration in liver affections, dyspepsia, and diseases of the stomach and bowels generally. Being diffusible in its operation, it produces a free circulation in the vessels on the surface of the body, accompanied by a gentle perspiration. It does not exhaust like drastic purges; still its action is more universal, and it may be often repeated, not merely with safety, but with great benefit; this becomes indispensably necessary in cases of long standing, for in them intense temporary impressions made by strong medicines seldom, if ever, do good, and tend to injure the Stamina of the Constitution.

It is not pretended that this medicine will cure all diseases; yet as a family medicine its effects are more safe and certain, than any other known to us, employed as a Cathartic. It is advised that except in ordinary cases, as in Dyspepsia, Sick Headache, and Intermittent and Remittent Fevers, that it be administered by a Physician, many of whom now, and from its extraordinary therapeutic effects, it is anticipated that all of them soon will recommend it. In chronic diseases, such as dyspepsia, ill-conditioned ulcers, hepatic and other cuticular affections; in hepatitis, or complaints of the liver, and in which it is a specific, performing all the functions of Calomel; in local congestions, and in the determination of blood to the head; in bilious cholics, and all complaints requiring an aperient, it produces the most salutary, without any constipating effects. In glandular and other concealed affections of the skin, it determines to the surface, and therefore relieves the patient. This medicine is the long sought desideratum, and to those exposed to our changeable climates, and particularly the fevers of the South, it is almost indispensable.

There are numerous attempts to imitate MILES' COMPOUND EXTRACT, but without effect. It is prepared by a peculiar process, the composition of which is not concealed from any regular Physician; in a climate like this, where the biliary constitutes the greater part of human diseases, the antidote to their malignity has happily been discovered. Indeed its diaphoretic effects are so gentle, mild and searching, that with their use the diseases of the South are disarmed of their terrors.

Application for Agencies, and for this medicine, may be made to THOMAS BELL Wholesale Commission Druggist, No. 22 COMMERCE STREET, Philadelphia.

N. B. This medicine may now be obtained at most of the respectable Drug Stores in this city, and so soon as arrangements can be made, from the principal Druggists in the United States.

Advertisement for Dr. Miles' Compound Extract of Tomato.
(The Historical Society of Pennsylvania.)

The *Philadelphia Botanic Sentinel* objected to the use of the tomato pill because of its supposed effective substitution for calomel.

> Are we to understand that it may be exhibited with propriety only in those cases in which, according to the doctrines of the schools, calomel is indicated? And is it to be given for the same purpose for which calomel would be administered? If it is, save us from its effects! If it cannot be used with safety in all cases of disease, we have been too long convinced that such medicines are injurious ever to resort to it as a medicine; and if it will subserve the same purposes and produce the same effect, then our suspicion of it would be turned to an aversion to it.[23]

PHELPS' COMPOUND TOMATO PILLS

While these debates were under way, more serious difficulties arose for Miles in the eastern states. In February 1838 he was informed that a Dr. Rowland's Compound Tomato Pills had been advertised. Miles asked his agents in New York, Boston, and Hartford for more information about Dr. Rowland. Shortly thereafter he received word that Hoadley, Phelps and Co. in New York had suspended the sale of his pills and was "even circulating calumnies respecting it." He directed Thomas Bell in Philadelphia to investigate these charges. Bell arrived in New York unannounced, only to find the principals out of the office, but he spoke with the company's bookkeeper. Bell's version of this conversation was as follows:

> 'Ah,' said I, 'you have the Tomato medicine; do you sell much of it?' 'No! we have sold but a few dozen.' 'How is the article approved of?' 'Why, we don't know anything about it ourselves, but the impression in regard to it is very unfavorable; it *salivates* people.' 'Oh,' said I; 'well, I see by this showbill you have here, that you have *another* Tomato medicine.' The showbill stuck up was 'Compound Tomato Pills, Blood Purifier, and Universal Panacea, prepared by G. R. Phelps. Price 37½ cents per box.' The clerk remarked, '*That medicine* we can recommend to be a first rate article.' 'Oh,' said I, 'it's prepared, I supposed, by Mr. Phelps, of this firm.' 'No,' he replied, 'by his brother!'[24]

George Phelps's brother was Guy R. Phelps. He was born in 1802 in Simsbury, Connecticut, and graduated in 1825 from Yale's Medical School, where he claimed to have become acquainted with the medical qualities of the *Solanum Lycopersicum.* He settled in New York to practice medicine and worked at Hoadley, Phelps and Co. He returned to Simsbury to practice medicine in 1827; he moved to Hartford a few years later. He was elected a fellow of the state medical society in 1836. In November 1837 Phelps began advertising Dr. Rowland's Compound Tomato Pills and the following March marketed Compound Tomato Pills. Advertisements for both pills appeared simultaneously in Hartford and Boston.[25] The name was subsequently expanded to Phelps' Compound Tomato Pills.

Miles believed that Phelps, in collaboration with Hoadley, Phelps and Co., was attempting to cash in on his promotional efforts for tomato medicine. Miles's pills had been advertised in Hartford before Phelps's pills had. The name Phelps' Compound Tomato Pills was similar to Miles' Compound Extract of Tomato, sold in the form of pills. Miles had developed hepatine, of which the contents were unknown. Phelps promoted tomatine, although his references to it were extremely few. All that is known is that it was described as an alkaloid principle recently "discovered in the *Solanum* class of plants, particularly in the Tomato," and said to be "an antidote to mercury." Another advertisement affirmed that it was an "alkaline substance, extracted from the Tomato Plant, with other vegetable substances which have been found to modify and diffuse its effects." Several advertisements stated that it was "one of the most important and useful discoveries ever made." Another said it was a "vegetable Principle, which was, after laborious investigation, discovered and first used as a medicine by the author of these pills." Like Miles, Phelps issued brochures, circulars, and show bills. Phelps's system of agents in Connecticut, New York, and Boston was similar to Miles's network in the west and south. Despite these similarities, the two pills were completely different in composition. Miles claimed that his pills were made from the tomato fruit; Phelps claimed that his were made from the leaves and stalks of the tomato plant.[26]

Despite this basic difference, the advertising claims for the pills were essentially the same. These claims, as strange as they may seem today, were imbedded in medical beliefs and practices current in the mid-nineteenth century. According to medical historian J. Worth Estes, the whole point of many cathartics, such as tomato medicine, and most drugs said to affect the bile was the notion that emptying the bowels of their contents and emptying the liver's bile into the intestines would help remove noxious substances wherever they were in the body.[27]

The Tomato Pill War

According to Thomas Bell, Hoadley, Phelps and Co. spread rumors that Miles's pills caused salivation, which was an indicator that calomel or some other mercurial product was present in them. In May 1838 Miles claimed that these rumors had been circulating in Cincinnati for several months and "at length reached the Eastern cities." He blamed two groups for generating them. The first was the medical faculty, of whose "jealousies manifested towards it" he was well aware and who feared "for the fate of their purses." The second was the "*secrete NOSTRUM makers*" who were "in greater danger from the popularity of this medicine." To squelch the rumors that his pills contained calomel, Miles offered a $1,000 reward to anyone who could prove that presently or at any time in the past they had contained that substance or any other form of mercury. Phelps also put up a reward of $1,000 for proof that his pills contained calomel, which suggested that his pills too had been the target of the rumormongers. As Phelps's advertisement appeared only a few times, it is unclear how seriously he took the rumors against his own pills. It was relatively easy to determine whether or not mercury was a component of the pills, and almost any chemist could have done so. No evidence has emerged to indicate that either pill contained calomel or any other mercurial component.[28]

In June 1838 Guy Phelps went into partnership with three other Hartford residents to promote and sell his tomato pills. This partnership placed Phelps in the executive position and made James Gilman the manufacturer of the pills. The two other partners put up $1,200 apiece to purchase new pill-making machines and help increase production and promotion of Phelps's tomato pills. Hoadley, Phelps and Co. launched a major promotion campaign on behalf of Phelps's pills in the summer of 1838. Advertisements in the *New York Journal of Commerce* testified as to the benefits of Phelps' Compound Tomato Pills. Mr. M. Belden of Hartford credited them with curing scrofula, which in the nineteenth century was viewed as a generalized debility that took multiple clinical forms and produced a wide variety of signs and symptoms. B. Bolles and C. Mann believed that the pills had cured derangements of their livers. Dr. Nathaniel Hosher of Hartford and Dr. George Palmer of New London divulged other miraculous successes. The *Exeter News-Letter* in New Hampshire alleged that Luther Stowell, who for twenty years had "never been one day without bandaging his leg, from the foot to the knee," had been cured after ingesting Phelps's pills. Newspapers and medical periodicals published articles praising them. The editors of the *Thomsonian Manual* stated that a friend of theirs "was completely salivated last week by taking two of Dr. Phelps' Tomato Pills." The *Boston Morning Post* claimed that the pills possessed unequaled power for "purifying the animal fluids, opening and cleansing the channels of circulation, and . . . giving strength and vigor to the nerves of organic life." The *New York Transcript* disclosed that Dr. Phelps "had brought into successful use a most invaluable medicine." Having adventured upon its use, they could "cordially recommend it to others." The *Hartford Patriot and Democrat* proclaimed that, unlike other speculations of the day, Phelps's pill "justly deserves the popularity and patronage it is receiving." This success may have led Miles to decide that Phelps and his Compound Tomato Pills "must be krushed."[29]

The war began with an anonymous "Communication" published in the *New York Journal of Commerce* on September 11, 1838, claiming that Miles's discovery was presented to the public only "after its virtues were tested by a *long and protracted* series of experiments." As Phelps conducted no such experiments, the article warned the public against "such reckless and sinister compositions," which endangered the health and life of its users. Phelps's pills were a "baseless imitation" of Miles's pills, the writer went on, and Guy Phelps was a "quack" and a "charlatan." Since it was published in New York, the real target of the attack was probably Hoadley, Phelps and Co., who, according to the communication, fabricated the pills "to *injure the reputation* of Dr. Miles' medicine by calumniously asserting that it contained *Calomel*, while they were at the same time his appointed, and professed to be his faithful agents in that city for its distribution."[30]

These charges caught Phelps off guard. Six days later he responded in the *New York Journal of Commerce,* reporting that the author of the communication was none other than William Bell, the brother of Miles's agent in Philadelphia. Phelps stated that he had been continuously engaged in his medical practice since 1825. During this time, he had explored the medicinal properties of vegetables, including the tomato plant, which he used in various forms in his practice. He claimed that he had never seen Miles's pills other than in their packaged form and that the composition of his own pills was entirely different from that of Miles's. He asserted that he was not a quack and offered a certificate, duly signed by the secretary of the Hartford County Medical Society, to prove his professional qualifications.[31]

This response did not satisfy Miles. On October 11 he published a letter in the *New-York Daily Whig* protesting that he had manufactured and advertised his pills almost a year before Phelps. He repeated the charges noted in Bell's letter and added others to the list. Miles threatened legal action to force Hoadley, Phelps and Co. to return the unused portion of their shipment of his pills, which the firm did. Strangely, the returned order was the same size as the original shipment. Miles felt that Hoadley, Phelps and Co. had never sold any pills in the first place, which meant either that their stories of salivation were false or that they had replaced the pills with ones that contained mercury in hopes of proving their claims. Miles claimed that it was improper for Hoadley, Phelps and Co. to become wholesale agents for two similar drugs. He charged that Phelps had deliberately sold his Compound Tomato Pills initially under the name of a nonexistent person, Dr. Rowland. Rowland just happened to be Phelps's middle name. Miles judged Phelps a counterfeiter. But not quite. Phelps's pills could not contain the active ingredient hepatine, as only Miles knew how to extract it from tomatoes. In addition, Miles did not believe that Phelps's pills even contained tomatoes. The price of Phelps's pills was considerably lower than the price charged by Miles (thirty-seven and one-half cents per small box versus fifty cents). Tomatoes cost more than Phelps was charging for his pills. Indeed, Phelps was selling so many pills that "all of the tomato production in Connecticut" would not have been enough to make them. Miles therefore concluded that Phelps's pills were made not of tomatoes but of some cheaper substance. Miles attempted to publish the article in the *New York Journal of Commerce* but was turned down. He paid for advertising space in the *New-York Daily Whig,* but after the initial publication, the editors refused to reprint it.[32]

Phelps refused to respond to the specific charges in Miles's communication, stating that Miles was "unjust and unmanly" and that he would "not notice any *such* attacks from the *same* or any *similar* source." A few months later Hoadley, Phelps and Co. issued a circular in which Miles's imputations were "either correctly explained or disproved."[33]

Miles was not content to let it rest. He sent copies of his communication to his agents nationwide, suggesting that they reprint it in local newspapers. Phelps was promptly attacked in Philadelphia, Cincinnati, and Baltimore newspapers even though he had not introduced his medicine into those places at that time. In Philadelphia the attack was particularly harsh. Phelps's actions were called unprincipled, and his pills were declared counterfeit. Phelps, it was charged, had published "the most **BAREFACED FALSEHOODS** in relation to his own spurious progeny." Only one falsehood was identified. Phelps had claimed that his tomato pills were patented. A letter from the U.S. Patent Office stated that Phelps had not applied for a patent. The advertisement containing this claim was signed by Abijah Freese, who identified himself as secretary of Miles's association of physicians. During subsequent months Phelps was "vilified in nearly all sections of the U.S." For instance, in Charleston, South Carolina, as soon as Phelps's tomato medicine was advertised, Miles's local distributor immediately reprinted the same article that had been used to attack Phelps in Philadelphia.[34]

An exception was in Hartford. Miles's agent there, Charles Hosmer, was ordered to attack Phelps but refused to do so. Miles promptly replaced Hosmer with Lorenzo Bull of Hartford, who was charged with taking a more aggressive posture toward Phelps. He began issuing advertisements warning the public about the unidentified imitations. Phelps countered by cautioning the public against unnamed attempts to deceive it with forgery. In early January Miles published a circular, "Medical Recorder," in which he attacked Phelps. In it he repeated charges that Phelps was a counterfeiter and claimed that Hoadley, Phelps and Co. had apologized for saying his pills contained calomel. Twenty-five thousand circulars were sent to postmasters around the nation and to individuals who gave testimonials on behalf of Phelps's pills.[35]

During the winter and spring of 1839 Phelps expanded his network of agents in New England, New York, and throughout the rest of the United States. In Pittsburgh Phelps selected as his agent H. Smyser, who began advertising Phelps's pills in late June. Smyser wrote to Phelps that

> there never was a medicine in this market, that every one extolled as they do your Tomato Pills. We have never had a Pill here that has given more uniform satisfaction; several of our most respectable citizens have used them, and express themselves very highly about their effects. *Miles' Tomato Extract* is no longer asked for, as your (Pills) have entirely monopolized the market.[36]

In Cincinnati, W. H. Harrison & Co. was selected as Phelps's agent. Harrison soon began an advertising campaign on behalf of "the only genuine *Compound Tomato Pills.*" In Pensacola, Florida, Phelps selected J. O. Smith as a retailer, and in Savannah, Georgia, G. R. Hendrickson, a gardener, sold them. In Charleston the druggists Haviland, Harral & Allen were selected as both wholesalers and retailers for Phelps's pills. They promptly proclaimed that the pills had received universal approbation for curing scrofula, dyspepsia, jaundice, bilious diseases, gravel, rheumatism, coughs, colds, influenza, catarrh, nervous diseases, acid stomachs, glandular swellings of all kinds, costiveness, colic, headaches, and many other diseases. It was also said that Phelps's pill was "an antidote to Contagious and Epidemic diseases," prevented "the formation of Bilious and Liver affections, Fever and Ague," and was the best cathartic for those who resided "in hot climates, and low and marshy countries." If you were a seaman, it was "an infallible remedy for the Scurvy." If you were a traveler, it was the best medicine "to counteract the dangers of exposure in unhealthy climates." As for ordinary family physic, Phelps's pills were universally approved and the best ever offered. If you ate too much at dinner, it was claimed, you could take some pills half an hour later and they would stimulate "the digestive powers of the stomach to a healthy and invigorated action."[37]

In May 1839 Guy Phelps found out that Miles intended to launch a major negative advertising campaign against Phelps' Compound Tomato Extract. He encouraged his brother to mount a promotional campaign on behalf of Phelps's pills. Guy Phelps approached his partners, who refused to authorize any expenditures for additional promotion. In exasperation, Guy Phelps wrote to George stating that his partners were narrow-minded and shortsighted. George Phelps was convinced that if he did not respond to Miles's attacks, within three months he would be unable to sell a pill. He told his brother that he almost "did not care if there were any pills sold or not."[38]

George Phelps's intelligence was correct. In June, Miles began a coordinated two-pronged offensive, one in Hartford and the other in New York City. In Hartford, Lorenzo Bull published an article from the *Xenia Free Press* in which he claimed that Miles and a company of physicians had obtained a medicine from the tomato that promised "to advance, in a very high degree, the health and safety of the human race." Phelps could not have been "ignorant of the previous discovery of Dr. Miles and others." According to the advertisement, Dr. Phelps's "right even to the *name of Tomato* for his medicine" was extremely doubtful.[39]

In New York, Miles launched his first assault with an advertisement in the *New York Transcript* on June 15. It included F. W. Chester's article, originally published in the *Cincinnati Journal and Luminary* in 1837, demonstrating that Miles's extract was the "original" tomato medicine. Chester, who then lived in New York, requested that Miles not use his name. Despite this request, Miles continued to publish Chester's article. The advertisement went further to stipulate that all other medicines "embracing the name tomato have no claim to the name" and were spurious articles developed by pretenders. This article was reprinted in Hartford during the following week.[40]

Finally, with the support of his partners, Phelps counterattacked. In Hartford he reprinted an article from the *Political Beacon* that was written by Dr. Robert Peters, professor of chemistry and pharmacy at Transylvania University in Kentucky. Peters had applied to become an agent to sell tomato pills. When he received his shipment, he analyzed their contents. His report stated that the white pills were composed of "aloe, rhubarb, pepper, colocynth, and some essential oil." Aloe was a drug with a nauseous odor, a bitter taste, and purgative qualities. Colocynth was a plant from the gourd family that produced a bitter, resin-like substance with purgative qualities and served as a harsh cathartic. According to Peters, the yellow pills contained rhubarb, pepper, aloe, and "some liquorice; and if any extract of Tomato, no essential quantity." Licorice and the root of the rhubarb were considered a cathartic. Peters returned the shipment of pills, believing them to be "an imposition upon the community," and stated that selling "a nostrum was derogatory to his standing as a gentleman or a man of science." As Phelps's pills came in only one kind and one color, which was neither white nor yellow, Phelps concluded that Peters had analyzed Miles's pills, which came in the two kinds and their corresponding two colors.[41] Lorenzo Bull denied that Peters had analyzed Miles's tomato pills and that the pills contained aloes or rhubarb. He charged, however, that Phelps's tomato pills were extensively believed to contain both aloes and rhubarb. Bull proposed that, as Miles had denied that his pills contained either ingredient, Phelps should therefore deny that his pills contained them as well. Phelps declined "to offer contempt and insult to an intelligent and discerning profession, by making the *proposed* denial." Phelps also wished to "avoid the degradation of having any sentiment *whatever 'put on record'* with those of the *Doctor* (?) Miles."

There was another reason Phelps did not wish to deny that his pills contained these ingredients. It was highly likely that they did. Although the formula for his pills has not been uncovered, his invoices indicate that he purchased gamboge, liquorice, aloes, alcohol, green fragment, gum arabic, cinnamon, and cayenne pepper. His invoices during this period made only one reference to tomatoes.[42]

Phelps claimed that Miles and his agents had filched the name of his product by using the name Tomato Pills, as Lorenzo Bull had been doing. Further, Phelps asserted that Miles's supposed tomato extract was really the American Hygiene pills that he originally made in Brunswick. In the words of an anonymous correspondent from Brunswick, Miles "never intimated that Tomato had any thing to do with them." Miles had taken several barrels of American Hygiene with him when he left Brunswick in the spring of 1837. By some mysterious process, his American Hygiene pills had been transformed miraculously into tomato pills. As Miles began advertising his tomato pills in June 1837, before the tomatoes had ripened, his pills could not have contained tomatoes as an ingredient. With respect to Miles's claim that he was a physician, Phelps published extracts of letters claiming that Miles had never attended medical school, that he was not a physician, and that "he had about as much claim to the title of doctor as my horse, and no more." Phelps pitied Miles's agents, who were in his opinion the unwitting and innocent victims of "a crafty, intriguing, and deceitful man." He also made a vague reference about judicial proceedings in which Miles had been involved. As the details of this case later emerged, this presumably was a warning to Miles to end the war, or the details of the case would be released.[43]

This time Miles and his agents were unprepared. In Hartford a pseudonymous letter, signed by "Urban," contradicted Lorenzo Bull's earlier statement about Professor Peters's analysis. Urban positively denied that Peters ever made a proper analysis of Miles' Compound Extract of Tomato. Peters made an attempt to analyze the contents, he said, "but who does not know that it is impossible to analyze with certainty a purely vegetable compound." Miles's medicine contained neither aloes nor rhubarb but certainly did contain tomatoes. While Phelps's pills were a mild cathartic, he continued, Miles's pills actually were a substitute for calomel. According to Urban, Miles had not produced more medicine in 1838 because the tomato cultivation throughout the country could not keep up with his needs. He facetiously stated, however, that "Dr. Phelps . . . found no difficulty here. O no! He could make Tomato pills enough to supply the world. Yea, Tomato plants were *so* abundant in this State in the summer of 1838, that *his* laboratory was in operation during the winter!"

The implication was that Phelps's pills contained no tomatoes. Urban branded Phelps the aggressor and claimed that he had information in his possession that cast doubt upon Phelps's medical credentials. He warned Phelps to cease his attacks or Phelps would regret it. Shortly after Peters's analysis appeared in print, Miles began publicly buying yellow tomatoes. He also advertised that yellow tomatoes were "doubly as valuable as the Red Tomato" and produced "twice as much hepatine, or active principle." As yellow tomatoes had but limited cultivation, Miles's purchases looked more like a public relations activity.[44]

Phelps promptly pointed out that the anonymity of Urban's letter suggested that the author was ashamed of his position or shrank "from the responsibility of his charges, in hiding himself from the public, and vilifying" him "most cowardly." Urban should "*have hoisted his flag*" with his "challenge to combat." Phelps claimed he had never been the aggressor, that he was only responding to Miles's attacks. With regard to his medical degree, Phelps stated that it had taken him longer to get his degree because of illness but that if Urban had anything else he wanted to report, he encouraged him to fire away. Urban, later identified as Isaac C. Pray, former editor of the *Bouquet,* was then a salaried agent for Miles in New York.[45]

Phelps also began a series titled "Piracy" composed of incidents in which Phelps claimed that others—particularly Miles and his associates—had attempted to steal his ideas or the good name of his tomato pills. Phelps claimed that he had located 968 instances of piracy, but only eight of them were published. One number included a letter from Dr. A. J. Holcombe stating that he, not Miles, was the real discoverer of tomato pills. Phelps concluded that Miles had no right to claim to be the originator of tomato medicine. Another number claimed that Phelps's name was used to sell Miles's pills in Vermont at the direction of Lorenzo Bull, who was said to be an honorable man.[46]

In New York, Phelps published advertisements repeating the charge that Miles's tomato pills were really a renamed version of his American Hygiene pills. Phelps claimed that Miles's advertisements contained "various insinuations and false representations" intended to impair the public's confidence in his pills. Miles's agents responded with an advertisement that essentially repeated his charges against Phelps. With regard to the veiled insinuations, Miles's agents taunted Phelps, "If Dr. P. knows any thing against Dr. Miles, let him make a *direct charge,* and the friends of the later will be sure to give the former an opportunity to show *proofs.*" On July 19 Miles's agents published a second article, "Read This and Decide," in which the priority for Miles's tomato medicine was again claimed, based upon published newspaper articles and advertisements. The advertisement also cautioned southern druggists against believing a copy of the *Northern Courier* dated April 25, which was a promotion for Phelps' pills. Evidently Phelps printed 15,000 copies of a version of the paper that was different from the original of that date, under the condition that it not be distributed in Hartford.[47]

Phelps attempted to respond by publishing a letter from an anonymous attorney in Cleveland reporting on a court case in which Miles was involved and an unsavory business deal that had gone sour. The attorney called Miles a "shrewd, arch, slippery fellow." The *New York Times* refused to publish the letter, but it was still circulated in New York. Isaac Pray responded with an article in the *New York Daily Express* on July 24 under the headline "Outrageous Libel." Pray stated that Phelps's attack, for "grossness and daring malignity, has seldom been equaled." With regard to the libelous letter, he said, Miles had paid every cent owed his creditors. He was a man of high respectability. He had been a member of the Ohio State Legislature and had been recommended as a judge "by a large number of the most respectable men in the town where he resided." The correspondent also divulged that "Dr. Miles has met with a serious domestic affliction which such a libel must have a tendency to aggravate." Phelps was challenged "to prove himself entitled to the benefits of the discovery—which he cannot prove and dare not attempt to do." Another communication, issued by Abijah Freese in the absence of Archibald Miles, stated that Phelps's advertisements contained "false and slanderous statements, based upon the authority of certain anonymous communications, manufactured for the purpose of injuring the reputation of Dr. Archibald Miles."[48]

Phelps, sensing victory, promptly went in for the kill. In a lengthy article published in the *New York Times* on August 1, 1839, he admitted his error with regard to the advertisement averring that he had patented his medicine. He claimed that he had considered applying for a patent, but after investigating, he was unwilling to disclose his process for making his pills. While he was considering patenting his pills, however, he published his first brochure with a statement that his pills were patented. Subsequently he applied for a copyright and removed the statement that his pills were patented. Since druggists advertised them as patented well into 1839, Phelps admitted that this was a careless mistake.[49]

Phelps had previously alluded to a court case in which Miles had been involved, and he now released the details. In Cleveland, Miles had been called a thief and had promptly sued for defamation of character but had lost the suit. He had received a letter from lawyers who had investigated the charges, and this time the *New York Times* agreed to publish extracts from the letter. He announced that the complete letter was available to anyone who wanted to see it. He also pointed out that Miles had created his association of physicians only after the Ohio legislature had turned down his application to create a "College of Health," which was intended to promote the selling of his pills. According to another correspondent, Miles had never mentioned his tomato medicine or had any interest in tomatoes prior to the spring of 1837, when he left Brunswick. Therefore, his claims that he had spent several years scientifically experimenting with tomatoes were without foundation. Miles had written a letter shortly after his arrival in Cincinnati, reporting that his pills were selling well under a new name. Another letter claimed that no physician in Brunswick used Miles's pills and none recommended them. Phelps therefore claimed that his were the only original tomato pills and that Miles was a fraud.[50]

Miles may have been wounded, but he was not out. He quickly gathered certificates from prominent individuals in Cincinnati, Brunswick, and Cleveland including members of the Ohio legislature, members of Congress, and mayors. All attested to his personal virtues and character. Isaac Pray published two columns of the testimonials, under the headline of "Calumny

Refuted," in the New York *Sunday Morning News* on September 1. With regard to Miles's interest in tomatoes, one correspondent claimed that Miles had told him as early as 1833 that he was experimenting with vegetable substances and that he had a particular interest in the tomato. The foremen for Miles's laboratories affirmed that large amounts of tomatoes had been shipped into them in 1836, 1837, and 1838.[51]

Lorenzo Bull reprinted the article in the *Hartford Daily Courant* on September 11 with a preface stating that Phelps had been the aggressor. Phelps responded the following day with an article titled "Pirates," in which he stated again that he was responding only in self-defence to Miles's attacks. He recommended that interested readers examine his defence previously published in the *New York Times.* Miles had stated to others that his persecution of Phelps "should not be mild, nor of short duration!!" Additional letters challenged Miles's credibility. One alleged that he had made use "of his christian character, as a means of carrying on his speculative designs; while he is a sly, cunning, insinuating, unprincipled man, and a *hypocrite of the deepest dye.*"[52]

Miles's agents responded with "Calumny Refuted *Again,*" with more certificates stressing the virtues of his character. Jacob Ward, a Methodist minister in Brunswick, affirmed that Miles "studied medicine in his youth, with the intention of becoming a practicing Physician, but declined it, for business pursuits." Phelps retorted that Jacob Ward's letter proved that Miles was not a physician. Miles was a merchant. Phelps ended with "Let the Painter stick to his pallet, and the Merchant to his desk." Miles had also told Ward that he was experimenting with vegetable substances as early as 1833. Phelps promptly published a column headed "The Poor Unfortunate Pirates, Great Excitement!" in which he stated that Bull had found it necessary to publish the long list of testimonials in order to confuse the public. All they proved was that there were a few men "in the whole state of Ohio" who did not "know of certain numerous and reprehensible transactions, of which the majority of the inhabitants" did know, and which Phelps could prove.[53]

Lorenzo Bull rebutted Phelps's account of Miles's court case in Cleveland. According to John Wiley, the former mayor of Cleveland and a former member of the Ohio senate, the defendant, Morris Seeley, had claimed he had never said that Miles was a thief. The decision had gone for the defendant because Miles could not prove that Seeley had made the statement.[54]

Phelps hit back, stating that Miles had not responded to many of his claims. Therefore the public should assume that Phelps's charges were correct: Miles was not a physician; his Extract was originally his American Hygiene pills; and Miles had been the aggressor from the beginning.

According to Phelps, Miles's "object was to get up an excitement, no matter how; *and if possible to get into a newspaper quarrel* or a law suit, and get a run of a year or two, and get rich which was all he wanted." Phelps presented additional excerpts from anonymous letters. One claimed that Miles had a brother in Brunswick who purchased "hundreds of pounds of *green mandrake roots,* which is ground there in a *Ring mill,* the juice expressed and boiled down, put into jars and sent off." Phelps asked, "Is not this the Hepatine?" Another proclaimed that he "*did* know a number of individuals who took a considerable quantity" of Miles's pills, "*but they are all dead.*"[55]

The tomato pill war ended abruptly in late 1839. Whether there was some agreement to end it or both sides simply determined that it was not in their financial interest to continue it is unknown. However, from October onward, only positive advertisements for their respective pills were published. Their advertising campaigns decreased appreciably as 1839 ended. In early 1840 both Miles and Phelps incorporated new strategies into their trimmed-down advertising campaigns. Miles's campaign stressed the medicinal virtues of the tomato, publishing numerous quotes that had appeared since Bennett had made his original comments in 1835. The excessive focus on this topic suggests that many people were questioning the basic premise of tomato medicine, which was that the tomato contained miracle ingredients. Phelps's campaign concentrated on particular groups. Some advertisements stated that Phelps's pills were particularly useful for women between the ages of twenty and forty. Others targeted ethnic and linguistic groups, publishing advertisements in French and Spanish.[56]

Only one tomato pill box is known to have survived. It is an oval wooden box two inches long and three-quarters of an inch wide with the name Phelps' Compound Tomato Pills on the top. A thin strip of wood connects the top to the bottom. When the box was manufactured, or how many pills it contained, is unknown. The box has the price of twenty-five cents on one side. Guy R. Phelps's signature extends the length of the other side. The box was constructed so that when it was opened, the signature was destroyed. This was evidently intended to prevent the box from being refilled with some other concoction by a counterfeiter.

The visibility and success of Miles's and Phelps's tomato pills encouraged others to develop and promote tomato medicines. Hallock's Tomato Panacea was produced and advertised in New York. It was billed as "a certain remedy for the following complaints dyspepsia, scrofula, or King's evil, rheumatism and nervous afflictions, worms, constipation, pulmonary and bilious complaints, eruptic diseases of the skin and all diseases arising from impure blood." Dr. Payne's Compound Tomato Pills were also available, as were other tomato preparations. Counterfeit tomato pills were produced with titles such as J. Phelps, W. Phelps, and Phelp's Tomato Pills. Phelps offered a $200 reward for finding the person who had counterfeited his boxes and his signature in Philadelphia and Pittsburgh. By the late 1830s panaceas and sanitary powders of all kinds said to be made from tomatoes were in high favor. According to the gardener Robert Buist, by 1840 "every variety of pill and panacea" was tomato extract. Tomato pills, tomato tinctures, and tomato decoctions were "conspicuously advertised in drug stores, with wreaths of crimson fruit placarded on the boxes."[57]

During the summer of 1839, almost every issue of every newspaper published in Hartford included at least one and in most cases numerous advertisements for tomato pills. Some issues contained as many as fifteen different advertisements that consumed more than thirty percent of the entire newspaper. Often advertisements were published for months and years on end. A correspondent to the *Connecticut Courant* postulated that "the Tomato Pill vendors intended to monopolize" the paper "to the exclusion of all other business men." Whether the subject was "of sufficient importance to justify this or not," tomato pills were "the engrossing topic of the day."[58] Thousands of advertisements for them have been located in newspapers throughout the United States, and probably tens of thousands were published.

The national market for tomato pills collapsed in 1840. Phelps ended his national advertising in the spring. Miles ended his in the summer. Miles's advertising in Cincinnati continued sporadically through the end of the year, and advertisements for Phelps's pills continued to be published in Connecticut and intermittently throughout New England for a few years. Phelps improved his pills in 1844 and advertised them under the generic name Tomato Pills. These improved pills, he claimed, could also cure jaundice, rheumatism, colds, heart palpitation, and lung fever in addition to the previously noted cures. Such promotions continued through early 1845.[59]

Despite the resulting visibility, the negative attacks by both Miles and Phelps could not have helped dispel the skepticism of many Americans toward such nostrums. A correspondent to the *Whig and Ægis,* reprinted in the *Boston Cultivator,* facetiously revealed that tomato pills "have a grand specific! They have a vast concentrating power capable of drawing both fools and coppers around them, if they could but be applied in a manner where their vast force might be exerted at once, many suppose they would draw the old and new worlds together and thus play the very devil with the big steam ships!"[60]

William Darlington, a physician and botanist, denounced the vendors of medical nostrums who had seized upon tomatoes "as a means of levying an additional tax upon the credulous." John Skinner had but little faith that tomato pills were a substitute for calomel. The *Western Reserve Magazine of Agriculture* decried that the tomato's healthful reputation had been employed to draw the public to the humbug known as tomato pills, which, the magazine's editors thought, were really composed of gamboge, a dense, reddish-yellow tree resin that had harsh cathartic results when taken internally. A silly jingle emerged: "Tomato pills, will cure all your ills."[61]

In spite of this criticism, the tomato was combined with other ingredients in pills throughout the 1840s. Ransom and Stevens, druggists in Boston, made Dandelion and Tomato Panacea in 1843 and declared that it surpassed "all other preparations in curing the worst of humors and eruptions of the skin." It never failed "in cases of indigestion, Dyspepsia, Loss of Appetite and Heartburn." It stood unrivaled in treating "Headache, Dizziness in the Head, and Jaundice Complaints." It was advertised only for a few months. In 1846 Frederick Brown, also a druggist in Boston, advertised Brown's Sarsaparilla & Tomato Bitters. It purportedly cured many of the same problems that earlier tomato medicines had been touted as healing and was also declared a certain cure for sinking of the stomach, lowness of spirits, hectic fever, night sweats, piles, and coughs. It received endorsements in the *Portland American,* the *Boston Daily Mail,* and the *Portland Bulletin.* It survived for a few years. In 1847 Dr. Warren's Sarsaparilla, Tomato and Wild Cherry Physical Bitters was similarly advertised, but like the others, it disappeared.[62]

Neither the claims nor the known contents of the pills nor their advertising campaigns were unique to tomato pills. During the early nineteenth century advertisements for thousands of nostrums, panaceas, and pills flooded America and purported to remedy bilious complaints. Many were the causes of this universal problem, including polluted water, bad food, eating too much, eating too fast, contagion, and "a diet stressing starchy dishes, salt-cured meats, fat-fried foods and lacking in fresh fruits and vegetables." Behind the claims for tomato pills was the notion that, while there were many forces that might upset indigestion, there was only one major cause: impure blood. The blood needed to be purified through purgation, it was believed. Many pills, including those produced by Miles and Phelps, were essentially laxatives. The ingredients aloes, colocynth, and gamboge, which were probably present in both pills, were purgatives. They were also common ingredients of other pills sold at the time, such as Benjamin Brandreth's Vegetable Universal Pills. Brandreth had arrived from England in 1835. Within four years he was reputed to be worth over $200,000 from selling his pills. His advertising budget alone reportedly reached the astronomical cost of $100,000 annually.[63]

Archibald Miles remained in Cincinnati for a few years and became a real estate agent. Before 1850 he returned to Cleveland. In the 1850 census Miles listed his occupation as physician. He owned $12,500 worth of real estate, which, while not close to the fortune amassed by Brandreth, was still a sizeable sum by the standards of the day. Whether the funds for the purchase of the real estate came from the sale of tomato pills or some other source is unknown. Later in the century a relative of his, Franklin Miles, founded Miles Laboratories in Elkhart, Indiana.

Guy Phelps continued to sell tomato pills and other medicines of his own creation, such as Phelps' Restorative. In 1845 he became fascinated with the life insurance business, then in its infancy in Connecticut. In the following year he became a founder and secretary of Connecticut Mutual Life Insurance Company. He later became its president. When he died in 1869 the firm had accumulated capital in excess of twenty-three million dollars. Despite his full-time interest in the insurance business, Phelps continued to sell tomato pills through the early 1850s. In 1866, years after he had stopped selling the pills, he received a letter from A. N. Albin from Pittsfield, which said in part,

> I have been out of health for some three or four months with deficient secretion of bile and compound derangements of the digestion Organs and general health; I get no relief from ordinary medicines but I am sure if I have some of your Tomato pills they would help me. Shall I be asking to[o] much of you if I ask you to send to me some of your pills[?] I have hesitated to write you because I know you do not wish to be troubled in that way—but I feel as though I could not get along without them—If you will send to me any quantity from one box to a gross—I shall feel greatly obliged and will if it is in my power do you or some one else as good a favor.[64]

Tomato pills did more than just enrich the pockets of their makers. Presuming that they did not contain mercury or other toxic substances, and to the extent that physicians did substitute them for calomel, public health was improved. Since calomel was toxic, in many cases it probably harmed patients more than the disease with which they were afflicted. To the extent that the pills contained tomato as a component, they may have contained concentrated doses of vitamins A and C, both of which were healthful. Finally, tomato pills were a major contributor to tomato mania, which swept through the land during the late 1830s and 1840s.

NOTES

1. *Painesville Telegraph,* November 21, 1834.

2. *Botanico-Medical Recorder,* 6 (August 11, 1838): 360; (April 7, 1838): 217–18; *Cincinnati Farmer and Mechanic,* 2 (July 30, 1834): 143.

3. John C. Bennett, *The History of the Saints; or An Exposé of Joe Smith and Mormonism* (Boston: Leland & Whiting, 1842), 11; *Painesville Telegraph,* November 21, 1834.

4. *Painesville Telegraph,* December 5, 1834; Andrew F. Smith, "Dr. John Cook Bennett's Tomato Campaign," *Old Northwest,* 16 (Spring 1992): 61–75; Smith, "'The Diploma Pedler;' Dr. John Cook Bennett and Christian College at New Albany, Indiana," *Indiana Magazine of History,* forthcoming.

5. Lobelia was the chief product promoted by Samuel Thompson and his followers. For information about the medicinal effects of rhubarb, see Clifford M. Foust, *Rhubarb, the Wondrous Drug* (Princeton, N.J.: Princeton University Press, 1992). For information about Dr. Townsend's Sarsaparilla and other nostrums, see Madge E. Pickard and R. Carlyle Buley, *The Midwest Pioneer; His Ills, Cures, & Doctors* (Crawfordsville, Ind.: R. E. Banta, 1945), 283.

6. *Maine Farmer,* 3 (August 21, 1835): 227; *American Farmer,* 2d ser., 2 (September 1, 1835): 142; *New England Farmer,* 14 (September 2, 1835): 62; *Genesee Farmer,* 5 (September 19, 1835): 304; *Tennessee Farmer,* 1 (September 1835): 172; *Cultivator,* 2 (September 1835): 102–3; *Southern Agriculturist,* 8 (October 1835): 557; *Farmers' Register,* 3 (November 1835): 349; Robley Dunglison, *Elements of Health* (Philadelphia: Carey, Lea & Blanchard, 1835), 300; *Ohio Farmer and Western Horticulturist,* 2 (August 1, 1835): 119; letter from Bennett, dated August 21, 1835, in the *Daily Cleveland Herald,* August 22, 1835.

7. *Botanico-Medical Recorder,* 4 (October 10, 1835): 13–14; *New York Transcript,* September 14, 1835.

8. *Erie Observer,* October 3, 1835; December 5, 1835.

9. *Cultivator,* 4 (June 1837): 62; *Southern Agriculturist,* 10 (April 1837): 190–92; Constantine S. Rafinesque, *The Pulmist or the Art to Cure and to Prevent the Consumption* (Philadelphia: C. Alexander, 1829), 50; *Botanico-Medical Recorder,* 6 (January 27, 1838): 136–37; Rafinesque to John Cook Bennett, as in the *Botanico-Medical Recorder,* 6 (November 18, 1837): 58–59.

10. *Genesee Farmer,* 6 (December 17, 1836): 402; *Hocking Valley Gazette,* as in the *Liberty Hall and Cincinnati Gazette,* October 5, 1837; *Botanico-Medical Recorder,* 6 (November 4, 1837): 39; (April 7, 1838): 218; Abraham

Reese, *The Cyclopedia or Universal Dictionary of Arts, Sciences, and Architecture* (Philadelphia: Samuel F. Bradford and Murray, Fairman, 1825), vol. 22.

11. *New York Farmer,* 4 (November 1831): 287; *Botanico-Medical Recorder,* 6 (January 13, 1838): 123.

12. Despite repeated discussions of Bennett's life at Nauvoo, the real reasons for his excommunication remain unclear. Specifically, he was charged with "teaching that illicit intercourse was condoned by Church leaders." Others have claimed that Bennett and Joseph Smith were competing for the affections of Sidney Rigdon's daughter. Still others suggest that a power struggle was under way and Bennett lost. *Times & Seasons,* 2 (May 1, 1841): 404; John C. Bennett, *The History of the Saints; or An Exposé of Joe Smith and Mormonism* (Boston: Leland & Whiting, 1842), 31, 301; Mary Audentia Smith Anderson, *Joseph Smith III and the Restoration* (Independence, Mo.: Herald House, 1952), 58; *Iowa Farmer and Horticulturist,* 4 (June 1856): 12.

13. *Painesville Telegraph,* November 21, 1834; *Green County Sentinel,* as in the *Botanico-Medical Recorder,* 6 (January 13, 1838): 123; *Thomsonian Recorder,* 4 (January 16, 1836): 113.

14. Andrew F. Smith, "The Amazing Archibald Miles and his Miracle Pills: Dr. Miles' Compound Extract of Tomato," *Queen City Heritage,* 50 (Summer 1992): 36–48.

15. *New York Times,* August 7, 1839; *New York Daily Whig,* October 11, 1838; *Philanthropist,* July 14–October 31, 1837.

16. *New York Times,* August 1, 1839; *Connecticut Courant,* October 5, 1839.

17. *Iowa Sun and Davenport & Rock Island News,* August 11–November 17, 1838.

18. *New York Times,* August 1, 1839; *Philanthropist,* July 14, 1837–March 10, 1840; *Western (Cincinnati) Christian Advocate,* 4 (September 8, 1837): 78; *Southern Agriculturist,* 10 (November 1837): 615; *Cincinnati Whig and Commercial Intelligencer,* November 30, 1837–April 13, 1838; *Daily Picayune,* December 21, 1838–February 16, 1839; *Charleston Mercury,* August 1, 1838–December 31, 1839; *Connecticut Courant,* January 29–April 7, 1838; *Baltimore Sun,* January 1, 1839; *Iowa Sun and Davenport & Rock Island News,* August 11–November 17, 1838.

19. *Western (Cincinnati) Christian Advocate,* 4 (September 8, 1837): 78; *Maine Farmer,* 5 (October 31, 1837): 301; *Southern Agriculturist,* 10 (November 1837): 615; *Catholic Telegraph,* 7 (January 24, 1838): 47; *Philanthropist,* as in the *New York Sunday Morning News,* June 23, 1839; *Cincinnati Journal and Western Luminary,* October 3, 1837.

20. *Troy (Ohio) Times,* October 10, 1838; *Louisville Daily Herald* and the *Jeffersonville Courier,* as in the *Connecticut Courant,* June 15, 1839.

21. *U.S. Gazette,* August 25, 1838; *Philanthropist,* October 2, 1838; January 2, 1838; *Alton Commercial Gazette,* April 9, 1839; *Daily Herald and Gazette,* April 27–October 12, 1839; *Circulating Business Directory* (Philadelphia: Morris's Xylographic Press, 1838), 97; *Iowa Sun and Davenport & Rock Island News,* August 11–November 17, 1838; *Hartford Daily Courant,* December 28, 1837; *Connecticut Courant,* December 30, 1837.

22. *Botanico-Medical Recorder,* 6 (March 24, 1838): 207; (June 30, 1838): 317.

23. *Philadelphia Botanic Sentinel,* 4 (March 15, 1838): 235.

24. *New-York Daily Whig,* October 11, 1838; *New York Sunday Morning News,* September 1, 1839.

25. *New York Journal of Commerce*, September 17, 1838; *New York Times,* August 2–13, 1839; *New York Sunday Morning News*, September 1, 1839; *Connecticut Courant,* February 24, 1838; *Connecticut Courant,* March 24, 1838; *U.S. Gazette,* November 6, 1838; *Boston Morning Post,* June 16, 1838; *Northern Courier,* June 21, 1838.

26. *Boston Morning Post,* June 16, 1838; *Pensacola Gazette,* December 1, 1838–February 16, 1839; *Charleston Mercury,* January 11–February 16, 1839; "Synopsis of Dr. Phelps' Compound Tomato Pills," cited in the *New-York Daily Whig,* October 18, 1838; *Hartford Daily Courant,* September 10, 1839.

27. Letter from J. Worth Estes, M.D., to the author, dated September 12, 1993.

28. *Philanthropist,* May 1, 1838; *Hartford Patriot and Democrat,* January 5, 1839; *Philanthropist,* October 2, 1838; *Sunday Morning News,* September 1, 1839.

29. Agreement to set up partnership, dated June 1, 1838, Connecticut Mutual Life Insurance Co.; *New York Journal of Commerce,* September 1–October 18, 1838; *Exeter News-Letter,* July 4, 1839; *Thomsonian Manual,* 5 (January 1, 1839): 63; *Boston Morning Post,* June 16, 1838; *New York Transcript,* as in *Patriot and Democrat,* January 19, 1839; *Hartford Patriot and Democrat,* as in *Cincinnati Daily Gazette,* September 3, 1839; *Connecticut Courant,* July 27, 1839; *Hartford Daily Courant,* September 26, 1839.

30. *New York Journal of Commerce,* September 11, 1838.

31. *New York Journal of Commerce,* September 17, 1838.

32. *New-York Daily Whig,* October 11, 1838; *New York Times,* August 1, 1839.

33. *New-York Daily Whig,* October 18, 1838; *Hartford Times,* October 5, 1839.

34. *Cincinnati Whig and Commercial Intelligencer,* October 29, 1838; *U.S. Gazette,* November 6, 1838; *Baltimore Sun,* December 1–28, 1838; *New York Times and Commercial Intelligencer,* August 1, 1839; *Charleston Mercury,* February 20–March 1, 1839.

35. *Hartford Times,* September 28, 1839; *Hartford Daily Courant,* September 28, 1839; *Hartford Patriot and Democrat,* December 15, 1838; January 5, 1839; Phelps's letter to Mr. Stone, dated March 19, 1839, and his letter to Hoadley, Phelps and Co., dated March 20, 1839, Phelps's Copy Book (April 1837–1839), Connecticut Mutual Life Insurance Co.; *New York Times,* August 1, 1839.

36. *Exeter News-Letter,* July 4, 1839; *Charleston Mercury,* January 11–February 16, 1839; *Ithaca Journal and General Advertiser,* January 7–June 16, 1839; *Troy Daily Whig,* February 25, 1839; Phelps's Copy Book (April 1837–1839), Connecticut Mutual Life Insurance Co.; *Daily Pittsburgh Gazette,* June 27–August 30, 1839; *Connecticut Courant,* July 13, 1839.

37. *Cincinnati Daily Gazette,* June 26, 1839; September 3, 1839; *Charleston Mercury,* January 11–February 16, 1839; *Pensacola Gazette,* December 1, 1838–February 16, 1839; *Savannah Republican,* June 4–12, 1839.

38. Phelps's letter to his brother, dated May 28, 1839, as in Phelps's Copy Book (April 1837–1839), Connecticut Mutual Life Insurance Co.

39. *Xenia Free Press,* as published in the Hartford *Daily Courant,* June 15, 1839.

40. *Connecticut Courant,* August 17, 1839; *New York Transcript,* June 15, 1839, as in the *Connecticut Courant,* June 29, 1839.

41. *Political Beacon,* September 29, 1838.

42. *Hartford Daily Courant,* July 4, 1839; Guy Phelps, Invoices, Connecticut Mutual Life Insurance Co.

43. *Hartford Daily Courant,* July 4, 1839; July 6, 1839.

44. *Hartford Daily Courant,* July 8, 1839; *Connecticut Courant,* July 13, 1839; *Cincinnati Chronicle,* July 16–27, 1839; *Augusta Chronicle,* June 18–October 31, 1839.

45. *Connecticut Courant,* July 13, 1839; September 14, 1839.

46. *Connecticut Courant,* July 20, 1839.

47. *New York Evening Post,* July 9, 1839; *New York Times,* July 15, 1839–August 1, 1839.

48. *New York Daily Express,* July 24, 1839; statement issued July 27, 1839, as published in the *New York Sunday Morning News,* September 1, 1839–January 5, 1840.

49. *Pensacola Gazette,* December 1, 1838–February 16, 1839.

50. *Connecticut Courant,* September 14, 1839; *New York Times,* August 1, 1839.

51. *New York Sunday Morning News,* September 1, 1839.

52. *Hartford Daily Courant,* September 11, 1839; September 12, 1839.

53. *Hartford Daily Courant,* September 18, 1839; *Connecticut Courant,* October 5, 1839.

54. *Hartford Daily Courant,* September 23, 1839.

55. *Hartford Daily Courant,* September 26, 1839.

56. *Connecticut Courant,* February 1, 1840; *New-England Weekly Review,* February 8, 1840; *Hartford Evening Courier,* January 24, 1840.

57. *New York Sun,* September 20, 1839; *Connecticut Courant,* November 30, 1839; July 20, 1839; *New York Express,* July 15, 1839; *Working Farmer,* 5 (January 1, 1853): 80; Robert Buist, *The Family Kitchen Gardener* (New York: J. C. Riker, 1847), 126; *Lancaster Farmer,* 11 (September 1879): 129.

58. *Connecticut Courant,* June 22, 1839.

59. *Cincinnati Daily Gazette,* August 17, 1840; *Connecticut Courant,* July 10, 1841; April 8, 1843; March 23, 1844; *New-England Weekly Review,* January 23, 1841; *Hartford Daily Courant,* January 29, 1845.

60. *Whig and Ægis,* as in the *Boston Cultivator,* 1 (September 21, 1839).

61. William Darlington, *Agricultural Botany: an Enumeration and Description of Useful Plants and Weeds* (Philadelphia: J. W. Moore, 1847), 132; *American Farmer,* 3d ser., 4 (August 17, 1842): 101; *Western Reserve Magazine of Agriculture and Horticulture,* 1 (July 1845): 100; *American Agriculturist,* 24 (October 1865): 318.

62. *Boston Post,* August 25–September 15, 1842; *Old Colony Memorial,* May 27–September 30, 1843; May 9–July 4, 1846; May 8–December 11, 1847; January 1, 1848; *Boston Daily Evening Transcript*, April 16, 1849–January 16, 1850.

63. James Harvey Young, *The Toadstool Millionaires: A Social History of Patent Medicines in America before Federal Regulation* (Princeton, N.J.: Princeton University Press, 1961), 76–79; Richard Osborn Cummings, *The American and His Food* (Chicago: University of Chicago Press, 1940), 4.

64. Oliver Seymour Phelps and Andrew T. Servin, compilers, *The Phelps Family of America* (Pittsfield, Mass.: Eagle, 1899), vol. 1, 617–20; *Obituary Record of Graduates of Yale College Deceased from July, 1859 to July, 1870* (New Haven, Conn.: Tuttle, Morehouse & Taylor, 1870), 329; Agreement with Richard Wood, dated October 28, 1852, Phelps Circular, dated June 1, 1856, and letter from A. N. Albin to Guy R. Phelps, dated February 7, 1866, Connecticut Mutual Life Insurance Co.

7

The Great Tomato Mania

Indications of an impending national tomato obsession were discernable during the 1820s. Cookbooks, gardeners' calendars, seed catalogues, almanacs, medical journals, agricultural periodicals, horticultural works, newspapers, and magazines published cookery recipes and directions for the cultivation of tomatoes. In parts of Florida, South Carolina, Virginia, and Louisiana tomato culture already flourished. In 1825 medical biographer Thomas Sewall reported that the tomato was raised in abundance in Virginia and the adjoining states and was regarded as a great luxury. In 1829 the editor of South Carolina's *Southern Agriculturist,* J. D. Legare, reported that the tomato was justly in high repute and was "a desirable object with most of our gardeners." Although there was little demand for the tomato in New England in 1830, a correspondent in the *New-England Palladium* proclaimed that it was a great favorite in Pennsylvania and New Jersey. The correspondent was so impressed with the tomato, he ended his comments with a flourish of Latin that had originally been said of sage: "Cur moriatur homo, cui crescit in horto. Tomatum?" (For those whose Latin might be a bit rusty, "Why might a man die, when tomatoes grow abundantly in his garden?") In 1831 Henry A. S. Dearborn, a politician and adjutant general in Massachusetts, heralded the tomato as a delicious and healthy vegetable. Of all vegetables, he believed, it was the most healthful, palatable, and refreshing.[1]

During the early 1830s, the number of published articles, recipes, and other references slowly escalated. The message was simple. The tomato was healthy, fashionable, and delicious. In 1831 Legare disclosed in the *Southern Agriculturist* that the tomato was a general favorite and was eagerly sought after. Luther Tucker, the editor of Rochester's *Genesee Farmer,* postulated that most people became fond of them after eating them a few times. Jesse Buel, the future editor of the *Albany Cultivator,* stated that tomatoes were highly conducive to health and one of the most desirable dishes upon the table.[2]

On October 23, 1831, Horatio Gates Spafford, a renowned New York writer, reported that ingesting tomatoes quickened the action of the liver and bowels, abolished headaches and "straitness" of the chest, and eliminated painful heaviness in the liver. Spafford regarded the tomato as "an invaluable article of diet, or, if you please, as of medicine, or of medical dietetics. With me, it has always been an object of solicitude, to find out such diet, as should supersede the necessity of medicine." Spafford's death shortly after the publication of his article did not discourage other journals and newspapers from reprinting his claims, which reverberated around the nation for years.[3]

The First Wave of Tomato Mania

Dr. John Cook Bennett's claims for the medicinal virtues of the tomato, first published in the fall of 1834, galvanized attention, generated excitement, and rapidly accelerated the acceptance of the tomato as a culinary vegetable. Over two hundred articles have been located that reprinted or paraphrased Bennett's claims, and probably many more were published. Some editors offered additional commentary. Southern newspapers, such as the *Alexandria (Virginia) Gazette,* Huntsville's *Southern Advocate,* and St. Augustine's *Florida Herald,* believed that Bennett's claims made "attributes to the *tomato* which will probably cause that delightful vegetable to be more generally used than it is at present." Without endorsing all of Bennett's conclusions, Jesse Buel in New York knew from his own experience that the tomato was salutary in the summer months. Buel noted that the tomato was "extensively used in the south and south-west, as an article of diet." The editor of the *Maine Farmer* could not say whether or not Bennett was correct in his statements in regard to the curative or preventive effects of tomatoes but acknowledged that they were "a useful article of diet, and should be found in every man's garden." The editor of the *New-England Palladium* believed that the tomato was a healthful vegetable, but he had a lingering suspicion that Bennett's claims were a hoax.[4]

In fact, almost everything Bennett had to say about the tomato was preposterous, but his wild and exaggerated claims made good press. Conversely, debunking Bennett's outlandish claims also made good press. Newspapers and periodicals ridiculed them for decades. Commenting on Bennett's claims, the editor of the *New York Transcript* sarcastically reported that

> A dandy in this city, who was troubled with that fashionable complaint, the dyspepsia, bought a dozen very large [tomatoes] in the market, which he presently stowed away in his breadbasket. It did the job for him, it must be allowed, for it cured dyspepsia and all other diseases, whether *in esse* or *in posse,* at the same time. Several other cases we have heard of, where the love apple has worked wonders in the stomach and intestines of its votaries.[5]

The *New York Evening Post* called Bennett's claims blarney and hoped that they would pass into oblivion. The *American Agriculturist* charged that his claims were libel on the medical profession and the tomato, concluding that "the whole thing savors of the most arrant quackery." Despite these charges, few editors resisted the urge to reprint them, as incredible as they were. American journals, newspapers, and cookbooks regularly repeated them for decades. Nor were Bennett's comments limited to publication in America. They swept across the Pacific and Atlantic oceans. The *Victorian Agricultural & Horticultural Gazette* published them in Australia. From Australia, they spread to London, where they were cited in British cookbooks, gardening publications, and medical journals in the mid-nineteenth century. A British gardening journal that published them in 1889 identified Bennett as an eminent physician. From the United Kingdom they were disseminated to the European continent, where they were cited as late as 1900.[6]

Whether or not they agreed with Bennett's claims, many editors and commentators seized the opportunity to extol the healthful virtues of the tomato. In Virginia a correspondent to the *Farmer's Register* certified that tomatoes had been tried by several individuals with decided good effect. The *Wheeling Times* stated that they were one of the greatest luxuries and were highly conducive to health. In Connecticut, Hartford's *Northern Courier* confirmed that tomatoes were an excellent remedy for the summer complaint. The editor of the *Baltimore Sun* doubted that the healthful effects of tomatoes were overrated and believed that the vegetable deserved far more general use, predicting that if "tomatoes, prepared with large quantities of stale bread and liberal use of salt, in the ordinary stewing mode were adopted as the food for children laboring under or recovering from 'summer disease,' the result would be highly gratifying."[7]

In New Jersey the *Morris County Whig* believed that Bennett's claims were "another inducement for their free use, besides the strong relish which most palates soon acquire for this delicious vegetable." In New York a correspondent reported in the *Genesee Farmer* that the tomato had been constantly used "in various forms, at almost every meal during the last three or four seasons, by myself and several acquaintance, whose health continued excellent even when the prevalence of cholera banished the fruits and vegetables generally from most tables."[8]

So much pro-tomato pressure was exerted that another writer in the *Genesee Farmer* stated that those who loved tomatoes should eat them but pleaded that they not impose them on others. The pro-tomato campaign, however, did not abate. In 1837 Jesse Buel welcomed "new evidence of the utility of this grateful garden vegetable in preserving health, and in curing indigestion, and diseases of the liver and the lungs." A correspondent in the *Lockport (New York) Democrat* reported that in

> all complaints attended with torpor and inactivity of the liver and bowels, it is a speedy and sovereign remedy; for sick headache, do.; for dyspepsia it never fails. For all calculous complaints, and diseases of the kidney and urinary passages, it is worth more, and will do more good than all other medicines put together, not excepting even Swaim's nostrum or Brandreth's *inert* pills. For scrofulous diseases generally of the glandular system, it is one of the best remedial agents that can be found, if not the very best. All diseases of an occult nature, requiring the aid of mercury in some form for their cure, will yield more readily to the free use of the tomato; and what is greatly in its favor, it does not, like mercury, ninety-nine times in a hundred, leave a more obstinate disease behind, and one more incurable than that sought to be removed. For the *modus operandi,* we must content ourselves by saying that the tomato acts as a detergent or wiper away, of crude and irritating humors floating in the system, and which frequently become entangled in the tortuous vessels of the glands, and there, if not removed, are sources of constant irritation and frequent disease. As a diuretic, it is not excelled by any known medicine; hence its efficacy in all dropsical affections. But time would fail me to enumerate all the diseases in which its virtues as a remedial agent have been tested.[9]

In the Midwest and South, pro-tomato commentary was just as persistent. The editors of the *Illinois State Register* believed there was "no vegetable that has been so grossly flattered" as the tomato. A correspondent from Charleston Neck commented in the *Southern Agriculturist* on the tomato's medicinal qualities, whose "salutary and benign influences upon all who use it" were "so apparent and perspicuous that the most ordinary observer can but appreciate them, as well as the most scientific practitioner of the healing art."[10]

The Second Wave of Tomato Mania

Bennett's claims might have faded into oblivion in 1837, but they were resuscitated by Alexander Miles and Guy R. Phelps. Bennett had predicted that a chemical would soon be extracted from tomatoes, "which will altogether supersede the use of Calomel in the cure of diseases." Miles, whose pills were purportedly made from the tomato fruit, directly cited Bennett's claims in support for his tomato pills. Phelps, whose pills were purportedly made out of the stalks and leaves of the tomato plant, also alluded to Bennett's claims. These claims laid the foundation for the massive advertising campaigns on behalf of tomato medicines, which led to the second wave of tomato mania.

Over six thousand advertisements for tomato pills have been located, and probably these were only a small fraction of the total number published. The advertisements not only promoted the pills but also made the case for the healthy tomato. While the quantity and content of the advertisements were amazing, hundreds of additional articles and recipes appeared in newspapers, magazines, cookbooks, almanacs, and periodicals. This promotional blitz surged through every region of the nation, rural and urban areas alike, and all Americans—lower, middle and upper classes—were infected with tomato mania.

The editors of the *Western (Cincinnati) Christian Advocate* announced that the tomato was highly medicinal and doubtless prevented many bilious attacks. F. W. Chester, the editor of the *Cincinnati Journal and Western Luminary,* asserted that the tomato was not only a delicious vegetable for the table but also a medicine believed to possess antibilious qualities. The *Liberty Hall and Cincinnati Gazette* recommended the tomato as a universal favorite that could be "brought to the table in varieties of preparation, not to be enumerated." Physicians agreed that they were not merely innoxious but healthful. Thirty-seven days later, the *Gazette* reported that tomatoes were beginning to make a voice in the world. The *Xenia Free Press* maintained that in Ohio the tomato was the very best and most efficient antibilious vegetable yet discovered. Its hepatic power was fully established both as food and medicine.[11]

Another physician prescribed the tomato for his own family and recommended it to his patients, who derived great benefit from it. Yet another stated that he had for years been recommending the *Tomato* as an article of refection, and particularly to his convalescing patients, as an excellent corrective in biliary derangement. Of particular interest was the small yellow tomato, which was in its season one of the greatest luxuries that could be brought to the table. Beside its excellence for eating, it was said to possess "great Medical virtue." Another observer commented that it was "well known that the Yellow Tomato contains abundantly, a Cathartic principle highly useful in Jaundice and other bilious diseases."[12]

In January 1838 Jesse Buel announced that there had been so much said in commendation of this vegetable as promotive of health that if he were to credit the declarations of the vendors, tomato medicine was an infallible cure for almost all diseases man was heir to. Ezekiel Holmes of the *Maine Farmer* asserted that tomatoes were highly esteemed by many as a table luxury. He was surprised that they were not more extensively cultivated since the tomato was an innocent luxury so easily and pleasantly afforded to him who would bestow a little pains and labor. The *New York Evening Star* stated that this healthy and most desirable vegetable was good for a cough and soothing to the lungs. It was recommended that it be used freely in hot months to check the accumulation of bile.[13]

Writers of seed catalogues, almanacs, and cookbooks joined in the praise. John Douglas, a gardener in Washington, D.C., proclaimed the "excellent and wholesome qualities of the tomato" in his seed catalogue. Dr. Page of Baltimore claimed that "Tomato Jelly" was excellent for all the pulmonary diseases. Catherine Esther Beecher, the sister of Harriet Beecher Stowe and Henry Ward Beecher, declared that the medical properties of the tomato were in high repute and that her recipe for "Tomato Syrup" retained all the healthful benefits of the fruit. Lettice Bryan, author of the *Kentucky Housewife,* placed her tomato soup recipe in the section of her book entitled "Preparations for the Sick" because she believed there was "no vegetable superior to the tomatoes, being very mild in taste, healthy, easily cultivated and yielding an abundant crop." In Massachusetts the botanist Chester Dewey explained that the tomato in many instances averted the evils of dyspepsia and kindred affections and was "a great disideratum for dyspeptics." Even federal government officials promoted the medicinal virtues of tomatoes. Commenting on a recipe for tomato figs, H. L. Ellsworth, commissioner of the U.S. Patent Office, exclaimed that "the medicinal qualities of tomatoes have greatly increased their cultivation, and every new preparation of the article is deserving consideration."[14]

MIRACULOUS TOMATO CURES

Along with the notion of miracle tomatoes came reports of miraculous cures. The *New England Farmer* announced that the free use of tomatoes was followed by rapid and permanent convalescence from disease of the liver.[15] Virginia's *Farmer's Register* advised farmers to serve tomatoes daily to slaves in the summer, as they proved an antidote to bilious fever. After ingesting tomatoes, persons were said to have recovered quickly from "chronic cough, the primary cause of which, in one case, was supposed to be the diseased liver—in another, diseased lungs." The editors of the *Yankee Farmer* asserted that

> One gentleman observed that he had been severely afflicted with dyspepsia for ten years—that he could not eat any boiled meat or vegetables, and often suffered much by this disorder; seeing the tomato recommended in the Yankee Farmer, last spring, as a good diet to correct this disorder, he cultivated the plant, and in the fall had a sauce made of the fruit. The first time he [ate] it, he did not like the taste, but after that he relished it well. He continued its use, and found it to be a pleasant food, and as a medicine it proved to be excellent, and afforded him relief from his complaint. He informed us a month or two ago, that since he used the tomato his health was better than it had been for ten years before.[16]

Followers of Samuel Thomson, the originator of Thomsonian medicine, presented the case of a lady from the western part of New York who "was afflicted with a general debility of the whole system, produced by a confirmed bilious habit and had also experienced three paralytic shocks, which had affected her left side and eye: her distress was excruciating." She was given a wineglassful of tomato syrup three times a day along with tomato pills at night. After two weeks, she left feeling much better. The improvement was attributed to the active properties of the tomato. Another physician described "two cases of inveterate disease of long standing (one of Consumption the other of Scrofula,) both considered hopeless, and both having been abandoned as incurable, [which] were cured by the extravagant use of Tomatoes for food."[17]

The *Boston Cultivator* remarked that "thousands of cases might be named of the tomato proving beneficial as a medical food, in restoring the system to a healthy tone when in a disordered state from various causes. Many of the most eminent physicians in the country bear testimony to its excellent medical effects under their direction and observation." The editor knew "a very severe case of dyspepsia, of ten years standing, cured by the use of the tomato. The patient had been unable to get any relief; he could eat no fresh meat, nor boiled vegetables. Reading an account of the virtues of the tomato, he raised some, and used them as food in the fall, stewed, and made some in a jelly for winter use. He was cured."[18]

During the mid-nineteenth century, cholera was a disease that frequently attacked American cities. Bennett and other physicians had alleged that tomatoes were a preventive against it. When cholera struck Cincinnati, tomatoes were an excellent remedy. According to the *Farmer's Monthly Visitor,* published in Concord, New Hampshire, a person in the last stage of the disease ate several tomatoes and revived. He ate more and "was soon able to return to the house with more strength than when he crawled out. He ate them every day, and recovered his health." In July 1849 Dr. Simon Pollak was stricken with cholera in St. Louis. The city council passed an ordinance forbidding that any kind of vegetables and fruit be brought to market or sold in the shops. Pollak's cravings for fresh tomatoes became intolerable. In spite of the strict prohibition of his doctors, he sent out to some of the truck gardens in the vicinity to purchase and smuggle into the hospital some ripe fresh tomatoes. He quickly consumed one tomato with great relish and waited experimentally but impatiently to learn its effect. After an hour he ate another and again another. When his doctors made their evening call he made a full confession. They were amazed and incredulous but did not censure him. To prevent his overdoing it, the doctors themselves ate up all the tomatoes on hand. Pollak recovered rapidly without another dose of medicine.[19]

While it is easy to dismiss the above stories, there may have been some truth in them. As has been previously mentioned, tomatoes contain significant levels of vitamins A and C. Mild deficiencies of vitamin A increase susceptibility to infection and promote abnormal functions of the gastrointestinal and respiratory tracts. Mild deficiencies of vitamin C lower resistance to infections, while severe deficiencies can result in hemorrhaging, anemia, and scurvy. Ingesting tomatoes may have improved the above conditions and strengthened a patient's ability to fight infection and disease.

THE TOMATO'S OTHER QUALITIES

It was not just the hoopla surrounding the medicinal tomato that caused the national preoccupation. Luther Tucker pointed out in the *Genesee Farmer* that even as a curiosity tomatoes deserved cultivation, as he did not "know of an annual plant of their size, that will produce so much fruit which, when ripened, from its beautiful red color, makes so pretty a show in the back ground." Even those who did not believe in tomato miracles believed the tomato to be a wholesome and delicious food. The editor of the *Boston Cultivator* believed that tomatoes had "no effect in cases of indigestion and complaints of the liver, though to others it may be beneficial." He regarded them "more as a wholesome food than a medicine" and thought their daily use was "harmless to the healthy." A correspondent in the *Southern Cultivator* was convinced that their general consumption would not only promote health but also improve the public's morals, although how this might be achieved was not spelled out. Mr. Milliken, the editor of Indiana's *South Bend Free Press,* wrote that tomatoes were "one of the most delicious and healthy vegetables that our country produces."[20]

The botanical qualities of the tomato plant contributed to its sustained success. It was easy to grow in almost every part of the United States and in many different soils and climatic conditions. In many areas tomatoes grew up unassisted in garbage dumps, weed patches, and other locations around habitations. They grew quickly and, depending upon the variety planted, could produce fruit within eighty to ninety days of planting. They were easily raised and a certain crop. New Jersey's *Morris County Whig* asserted that there was no vegetable that could be "cultivated with less care and trouble, and none which will, if properly treated, yield in greater abundance." Several sources reported that there was "no vegetable more easily raised, and none better pay the cultivator where they are generally known." In 1835 the editor of the *New England Farmer* announced that the tomato was "of recent culture in this part of the country." It was then "becoming popular as an article of diet, on account of its being esteemed salutary in cases of dyspepsia, &c. and as wholesome as it is palatable." He happily observed that the tomato was "rapidly coming into notice." Hartford's *Silk Culturist* advised farmers to "immediately take measures to ensure an extensive cultivation of this article the ensuing year." There was no vegetable that could "be raised with less trouble or in greater quantity than the Tomato," and there was none that paid the gardener "half so well for his labor." In 1844 the *Albany Cultivator* pointed out that tomatoes were "now much more common than formerly" because of the mandates of imperious fashion.[21]

The tomato was easier to harvest than were other vegetables such as the potato. In 1842 the editor of the *Boston Plough* posited that as many tomatoes could be raised on an acre as potatoes. As tomatoes did not have to be dug up, gathering them was done with far less labor than gathering potatoes. Likewise, during the 1850s, New Jersey tomato grower Edmund Morris preferred raising tomatoes at thirty-seven cents a bushel to potatoes at seventy-five cents because he did not have to dig into the ground for the tomatoes. Also, surpassing the expectations expressed some years earlier in the *Boston Plough,* the quantity that could be realized from an acre of tomatoes far exceeded that of potatoes.[22]

THE EFFECTS OF TOMATO MANIA

Tomato mania lasted only a few years, but its effects continued for decades. The sale of tomato seeds dramatically increased during the 1830s and 1840s. The *New England Cultivator* observed that in 1833 the only seed store in Boston was thought to be doing a remarkable business if it sold one pound of tomato seed a year. By 1839 a single store in Boston, Hovey & Co., had a hundred bushels of seeds for sale. By 1851 more than a thousand pounds of seed were sold in Boston alone.[23]

The sale of fresh tomatoes increased, as did the profits for market gardeners. In the early 1830s the editor of Boston's *American Gardener's Magazine* found tomatoes in the markets in abundance that could be procured at almost any price, even in the early part of the season. As the knowledge of the tomato's value became generally known, consumer demand for them increased. By 1835 tomatoes could be bought in the first part of the season only at a high price, which well repaid "the gardener for his trouble in growing them throughout the whole summer." In early 1836 the *New York Times* published a report claiming that "one gentleman last year cleared $1,800 by rearing this article on a small farm." This was a small fortune by the standards of the day. In 1838 Hartford's *Silk Culturist* announced that the investigations of medical men had greatly increased the consumption of the tomato and its price had doubled. In 1839 Edward Hooper, the editor of the *Western Farmer and Gardener,* maintained that a farmer near Cincinnati made $1,000 in one summer by raising tomatoes. In 1842 the *Boston Plough* reported that this fashionable vegetable commanded a price of three dollars per bushel. A correspondent in the *Farmer's Register* reported that a tomato patch of four hundred square yards produced a bushel of tomatoes a day during the tomato season. From a single acre of tomatoes, a farmer near Richmond gathered eight bushels of tomatoes a day, beginning in July and ending when the frost arrived in the fall. Even these figures paled by comparison with the estimates of profits in California during the Gold Rush, where one and a half acres of tomatoes purportedly sold for $18,000.[24]

During the 1840s and early 1850s, farms around large cities started to cultivate tomatoes extensively. The North American Phalanx in Monmouth, New Jersey, sent fresh tomatoes to New York beginning in the late 1840s. In July 1852 its first bushel of tomatoes sold for eight dollars. In 1854 Russell Trall reported that tomatoes at the first of the season sold for fifty cents a quart in New York. Edmund Morris, who began farming in 1855 on a ten-acre farm in southern New Jersey, shipped tomatoes to markets in both Philadelphia and New York. On one acre he grew 3,760 tomato plants, which produced 501 bushels of fruit and netted a yield of $190. Morris noted that older hands at the business did much better, occasionally generating as much as $400 an acre. He remarked that "a smart man will gather from sixty to seventy bushels a day." He estimated that the expenses of cultivating tomatoes ran about sixty dollars an acre, and the gross yield might be safely calculated at $250, leaving close to $200 in profit. By the standards of the 1850s, that was an excellent financial return. Morris stated that "if it were not for the sudden and tremendous fall in prices to which tomatoes are subject soon after they come into market, growers might become rich in a few years."[25]

Tomatoes were picked green and transported great distances. From southern states they were soon shipped to northern states weeks before they were available locally. During the late 1840s truck farmers from Maryland to Georgia hauled tomatoes to Philadelphia, New York, and Boston. This was such a lucrative trade that during the 1850s farmers from New Jersey emigrated to Virginia to grow tomatoes for export to northern states. Although this competition seriously interfered with the profits of New Jersey tomato farmers, it did not destroy them. Edmund Morris contended that when prices fell during the summer,

> the Southern growers could not afford the cost of delivery here, and thus left us in undisputed possession of the market. But, as a general rule, the Virginia competitors invariably obtained the highest prices. A great portion of their several crops, however, perished on their hands; because, as they had no market here when prices fell, so the scanty population around them afforded none at home.[26]

Tomato cultivation slowly increased during the early nineteenth century. During the mid-1830s, however, it sharply escalated. In early 1836 the *New York Times* predicted that the "consumption will be quadruple the next year what it was the year preceding." In 1839 the *Xenia Free Press* reported that in Ohio the tomato was "produced in vast abundance." In 1840 the *Pittsburgh Intelligencer* estimated that there were 50,000 bushels of tomatoes growing within a ten-mile radius of the city, with a single farmer growing five acres. Fearing Burr, the Massachusetts horticulturist, estimated that during the 1840s and 1850s tomato cultivation quadrupled. Robert Buist, the Philadelphia market gardener and popular author, guessed in 1847 that it occupied "as great a surface of ground as Cabbage."[27]

Tomato seeds vary in size and weight. Today they range from 8,800 per ounce for the beefsteak tomato to more than 21,000 per ounce for the cherry tomato variety. During the early nineteenth century tomato seeds averaged about 4,000 per ounce because of inadequate ways of cleaning them. If a thousand pounds of seed were sold in Boston, over sixty-four million seeds were sold. Obviously, not all seeds were planted, or, if they were planted, survived to bear fruit. Even if only a fraction of the seeds bore fruit, however, millions of bushels of tomatoes were grown in New England in the early 1850s.

By 1851 Patrick Neill affirmed that tomatoes were an article of immense cultivation in the southern and middle states. Around Philadelphia the tomato was considered the "prince of the vegetable market" and was "an object of extreme field cultivation." Edmund Morris reported that a vast area was planted with tomatoes. By 1858 thousands of acres were cultivated to supply the demands of large cities.[28]

TOMATO CONSUMPTION

What became increasingly clear as the 1830s progressed was that the extreme flexibility and versatility of the culinary tomato greatly enhanced its value. It was served at breakfast, lunch, and dinner. It was served in soups and salads, as main courses and as side dishes. It was recommended for use with eggs, all types of meats, poultry, and fish. It was made into sauces, ketchups, soys, relishes, and many other condiments. It was employed to make pastries, pies, tarts, marmalades, jams, jellies, and many other desserts and sweets. And of course, when there was a glut during the high season, it could be fed to cows, pigs, and other stock. There was no other fruit or vegetable that could boast such a wide array of uses. In 1831 a correspondent to the *Genesee Farmer* professed that "either raw or stewed—in soups, or fricassees—for gravy or catsup—as well as for pickles and sweetmeats—its utility is such that it would not readily be dispensed with by those who have given it a fair trial in these various ways." His several years of experience with the tomato enabled him to recommend it "to all who desire the acquisition in their gardens of a cheap luxury. For salubrity, none can surpass it." In 1835 the editor of the *New England Farmer* asserted that tomatoes were a dietetic luxury whose utility was "so great and so varied, that few who have once adopted its use, can be prevailed upon to dispense with it." In the same year, Luther Tucker expressed his belief that the tomato was not only very delicious but a very harmless and wholesome vegetable. Some gave "a decided preference to a dish of *tomato sauce* or a tomato pie, when properly prepared, to any thing of the kind in the vegetable kingdom." Tucker's experience in raising and using this vegetable in various ways enabled him to recommend it to all who were desirous of obtaining a cheap and delicious fruit for the table. The *Cambridge Chronicle* averred that "no person will be without the article after he becomes acquainted with its virtues, and accustomed to its use."[29]

In 1841 the Massachusetts horticulturist J. W. Russell stated that the tomato was an indispensable dish for every table through the summer months. The modes of cooking tomatoes varied according to the taste and fancy of individuals. The *American Farmer* announced that tomatoes were fashionable. The following year, the same journal proclaimed that "the celebrity of this plant has been astonishing. A few years since, prejudice reviled at its excellences with its most vindictive tauntings. Now, it is an article of so general popularity, scarcely a garden, or an apology for one, is to be found where it is not cultivated, and almost every voice is loud in proclaiming its excellences."[30]

Such consumption was not limited to the eastern and southern states. In Madison, Wisconsin, then on the frontier, a correspondent to the *London Observer* described a tomato feast while staying at the American Hotel in 1841:

> Tomato was the word—the theme—the song, from morning till night—from night till morning. The first morning I descended to the bar, there sat the colonel . . . his expressive mouth full of a red tomato. That swallowed, he held up another love-apple tantalizingly, to a feeble little child, and, mincing his voice, he would exclaim, "Who'll have a tomato? Who'll kiss me for a tomato?" In truth, not I; having in the early part of my days looked upon that grovelling fruit as poison, and never having tasted it even as a pickle with much gusto, I was not prepared to enjoy the tomato feast, at the capital of Wisconsin.
>
> The garden at the rear of the house seemed to produce no other fruit or vegetable. At breakfast we had five or six plates of the scarlet fruit pompously paraded and eagerly devoured, with hearty commendations, by the guests. Some eat them with milk, others with vinegar and mustard, some with sugar and molasses. I essayed to follow suit, and was very near refunding the rest of my breakfast upon the table, the sickly flavour of the flat-tongue grass, sour milk, and raw cabbage, being concealed under the beautiful skin of the love-apple I had the temerity to swallow.
>
> At dinner, tomatoes *encore,* in pies and patties, mashed in side dishes, then dried in the sun like figs; at tea, tomato conserves, and preserved in maple sugar; and to crown the whole, the good lady of the hostel launched forth at night into the praise of tomato pills.[31]

In 1842 the editors of the *American Agriculturist* pointed out that tomatoes were "used in various ways, either raw, with sugar, or stewed for sauce, or in a fricassees and soups; for catsup or gravy, for meat and for pies or preserves, as well as for pickles and sweet-meats." The *Cultivator* announced that all who had not deliberately made up their minds to be ranked among the "nobodies" had learned to eat tomatoes. A correspondent to the *American Farmer* admitted eating "this vegetable raw, without anything—cut up with vinegar, salt, pepper and mustard." He also ate tomatoes "fried in butter, and in lard, broiled and basted with butter, stewed with, and without bread, with cream and with butter." He said with a clear conscience that he liked them in every way they had ever been fixed for his palate.[32]

The *Connecticut Farmer's Gazette* reported that the tomato, against which a few years earlier there had been almost a universal prejudice, was now decidedly popular with the friends of good living and justly ranked among the choice products of the garden. Philadelphia's *Farmer's Cabinet* decreed that no farmer should be without tomatoes. The farmer's family, it said, "will soon want their tomatoes, —once, —twice, —three times a day, —morning, noon, and evening!" Henry Ward Beecher, editor of the *Indiana Farmer,* announced that whoever did not love tomatoes was an object of pity. By the mid-1840s tomatoes were cultivated the length and breadth of the country, in almost every garden from Boston to New Orleans, and were universal favorites from the Gulf to the Great Lakes. They were served on every table through their growing season. By the 1850s Professor Mapes, editor of the *Working Farmer,* calculated that from July to October the tomato was sold in larger quantities than any other vegetable. As a British visitor observed in the early 1850s, tomatoes had become the "sine qua non of American existence."[33]

NOTES

1. Thomas Sewall, *A Lecture Delivered at the Opening of the Medical Department of the Columbian College, in the District of Columbia* (Washington: Printed at the Columbian Office, March 30, 1825), 61–62; *Southern Agriculturist,* 2 (February 1829): 79; *New-England Palladium,* July 20, 1830; *New England Farmer,* 10 (December 21, 1831): 179.

2. *Genesee Farmer,* 1 (July 1831): 233; (August 1831): 266; *Southern Agriculturist,* 4 (February 1831): 81–82.

3. *Northern Farmer and Practical Horticulturist,* 2 (December 28, 1833): 73; *American Farmer,* 2d ser., 2 (September 2, 1834): 143; *Ohio Repository,* August 22, 1834; *Xenia Free Press,* August 16, 1834; *Logansport Canal Telegraph,* September 13, 1834; *Thomsonian Recorder,* 3 (October 11, 1834):

16; *Genesee Farmer,* 6 (December 17, 1836): 402; *Southern Agriculturist,* 10 (April 1837): 190–92; 2d ser., 1 (December 1841): 651-2; *New England Farmer,* 15 (April 26, 1837): 329; *Thomsonian Manual,* 4 (October 1, 1838): 175; *Connecticut Courant,* February 1, 1840.

4. *Alexandria Gazette,* August 8, 1835; *Florida Herald,* August 20, 1835; *Cultivator,* 2 (September 1835): 102–3; *Maine Farmer,* 3 (October 16, 1835): 289; *New-England Palladium,* September 25, 1835.

5. *New York Transcript,* September 14, 1835.

6. *New York Evening Post,* August 2, 1859; *Victorian Agricultural & Horticultural Gazette,* 3 (August 21, 1859): 81; *Australian Horticultural Magazine,* 1 (September 1877): 203–8; *Gardener's Chronicle,* November 19, 1859, p. 932; Isabella Beeton, *Book of Household Management* (London: S. O. Beeton, 1861), 1096; *Lancet,* as in *Garden,* 2 (November 16, 1872): 435; *Horticultural Times and Covent Gardens Gazette,* 5 (July 9, 1887): 26; William Iggulden, *The Tomato* (London: Journal of Horticulture Office, 1889), 12; *La Semaine Horticole,* as in *Meehan's Monthly,* 10 (May 1900): 77.

7. *Farmer's Register,* 5 (January 1837): 13–14; *Wheeling Times,* as in the *Liberty Hall and Cincinnati Gazette,* August 29, 1837; *Northern Courier,* September 6, 1838; *Baltimore Sun,* August 3, 1839.

8. *Morris County Whig,* August 12, 1835; *Genesee Farmer,* 1 (July 30, 1831): 233; 5 (January 24, 1835): 29.

9. *Genesee Farmer,* 5 (March 7, 1835): 78; *Lockport Democrat,* as in the *Southern Agriculturist,* 10 (November 1837): 615.

10. *Illinois State Register,* September 14, 1839; *Southern Agriculturist,* 10 (April 1837): 190–92.

11. *Western (Cincinnati) Christian Advocate,* 4 (September 8, 1837): 78; *Cincinnati Journal and Western Luminary,* October 3, 1837; *Liberty Hall and Cincinnati Gazette,* August 29, 1837; October 5, 1837; *Western Farmers Almanac* (Stubenville, Ohio: James Turnbull, 1839), 41–42; *Xenia Free Press,* March 23, 1839.

12. *Connecticut Courant,* February 1, 1840.

13. *Cultivator,* 5 (January 1838): 184; *Maine Farmer,* 5 (May 23, 1837): 115; 3 (October 16, 1835): 289.

14. Ellis & Bosson, *Annual Catalogue of Vegetable, Tree, Flower, Field and Grass Seeds* (Boston: Yankee Farmer Office, 1839), 20; John Douglas, *Catalogue of Kitchen, Garden Herb, Field and Grass Seeds* (Washington, D.C.: Alexander and Barnard, 1843), 32; *American Farmer,* 3d ser., 3 (October 7, 1840): 155; 3d ser., 3 (August 1841): 97; Catherine Esther Beecher, *Miss Beecher's Domestic Receipt Book* (New York: Harper & Brothers, 1846), 197; Lettice Bryan, *The Kentucky Housewife* (Cincinnati: Shepard & Stearns, 1841),

24–25; Chester Dewey, *Report on the Herbaceous Plants and on the Quadrupeds of Massachusetts* (Cambridge: Folsom, Wells, and Thurston, 1840), 166.

15. *New England Farmer,* 16 (May 16, 1838): 357.

16. *Yankee Farmer,* 5 (April 20, 1839): 122–23.

17. Samuel Thomson, *The Thomsonian Materia Medica, or Botanic Family Physician* (Albany, N.Y.: J. Munsell, 1841), 655–56; *Connecticut Courant,* February 1, 1840; *Valley Farmer,* 2 (August 1850): 241; John L. Blake, *The Farmer's Every-day Book* (Auburn, N.Y.: Derby, Miller and Company, 1851), 558.

18. *Boston Cultivator,* 5 (September 9, 1843): 282.

19. *Farmer's Monthly Visitor,* 1 (October 20, 1839): 147; Frank Lutz, ed., "Autobiography and Reminiscences of S. Pollak M. D.," in the *St. Louis Medical Review* (1904): 59.

20. *Genesee Farmer,* 1 (August 27, 1831): 266; *Boston Cultivator,* 5 (September 9, 1843): 282; *Southern Cultivator,* 11 (August 1853): 242; *Maine Farmer,* 23 (August 9, 1855); *South Bend Free Press,* August 28, 1838.

21. *Genesee Farmer,* 1 (August 27, 1831): 266; 5 (September 5, 1835): 282–83; *Cultivator,* 2 (September 1835): 102–3; 2d ser., 1 (March 1844): 100: *Morris County Whig*, August 12, 1835; *New England Farmer*, 10 (March 1832): 285; 14 (October 14, 1835): 106; 16 (May 16, 1838): 357; *Southern Agriculturist,* 12 (May 1839): 274–75; *Silk Culturist,* 4 (April 1838): 7.

22. Edmund Morris, *Ten Acres Enough* (New York: J. Miller, 1864), 118, 156–57.; *Boston Plough,* as in the *American Farmer,* 3rd ser., 4 (September 1, 1842): 124.

23. *New England Cultivator,* 1 (September 1852): 258; 1 (September 1852): 258; *Boston Cultivator,* 1 (January 19, 1839).

24. *American Gardener's Magazine,* 1 (February 1835): 45; *New York Times,* February 9, 1836; *Erie Observer,* February 27, 1836; *Silk Culturist,* 4 (April 1838): 7; Edward James Hooper, *The Practical Farmer, Gardener and Housewife* (Cincinnati, Ohio: Geo. Conclin, 1839), 493–94; *Boston Plough,* as in the *American Farmer,* 3rd ser., 4 (September 1, 1842): 124; *Farmer's Register,* 5 (January 1837): 13–14; *L. C. Advocate,* as in the *Southern Planter,* 3 (April 1843): 94–95; Edward Douglas Branch, *Westward: The Romance of the American Frontier* (New York: D. Appleton, 1930), 487; Fred A. Shannon, *The Economic History of the United States: The Farmer's Last Frontier; Agriculture, 1860–1897* (New York: Holt, Rinehart and Winston, 1961), vol. 5, 28.

25. A. J. McDonald, as in John H. Noyes, *History of American Socialism* (Philadelphia: J. B. Lippincott, 1870), 483; Russell Thatcher Trall, *The New Hydropathic Cook-book* (New York: Fowlers and Wells, 1854), 76; Morris, *Ten Acres,* 118, 156–57.

26. James C. Bonner, *A History of Georgia Agriculture 1732–1860* (Athens: University of Georgia Press, 1964), 169; Fearing Burr, *The Field and Garden Vegetables of America* (Boston: Crosby and Nichols, 1863), 643; Morris, *Ten Acres,* 118–19.

27. *New York Times,* February 9, 1836; *Erie Observer,* February 27, 1836; Hooper, *Practical Farmer,* 493–94; *Xenia Free Press,* March 23, 1839; *Pittsburgh Intelligencer,* August 22, 1840, as in the *New York Evening Star,* August 27, 1840; Francis S. Holmes, *The Southern Farmer and Market Gardener* (Charleston: Burges & James, 1842), 86–87, 96, 100, 105, 236; Burr, *Field and Garden Vegetables,* 639; Robert Buist, *The Family Kitchen Gardener* (New York: J. C. Riker, 1847), 128.

28. Patrick Neill, *The Fruit, Flower and Kitchen Garden* (Philadelphia: Henry Carey Baird, 1851), 236–37; *Philadelphia Florist,* 1 (May 1852): 30; Morris, *Ten Acres,* 128; *Southern Planter,* 18 (February 1858): 89.

29. *Genesee Farmer,* 1 (August 27, 1831): 266; (July 30, 1831): 233; 5 (January 24, 1835): 29; (September 5, 1835): 282–83; *New England Farmer,* 14 (October 14, 1835): 106; *Cambridge Chronicle*, September 26, 1835.

30. *Magazine of Horticulture,* 7 (March 1841): 97; *American Farmer,* 3d ser., 4 (September 1842): 124; 3d ser., 3 (May 1842): 414.

31. *London Observer,* as in *Life in the West: Back-Wood Leaves and Prairie Flowers* (London: Sounders and Otley, 1842), 266–67.

32. *American Agriculturist,* 1 (June 1842): 90–91; *Cultivator,* 9 (October 1842): 167; *American Farmer,* 3d ser., 4 (August 17, 1842): 101.

33. *Connecticut Farmer's Gazette,* 3 (April 15, 1843): 236; *Farmer's Cabinet,* 8 (August 1843): 44; *Cultivator,* 2d ser., 2 (October 1845): 321; Buist, *Kitchen Gardener* (New York: J. C. Riker, 1847), 126; Daniel Drake, *A Systematic Treatise, Historical, Etiological and Practical, on the Principal Diseases of the Interior Valley of North America* (Cincinnati: Winthrop B. Smith, 1850), 656; *Working Farmer,* 5 (June 1, 1853): 80; Nina Cust, *Wanderers: Episodes from the Travels of Lady Emmeline Stuart-Wortley and Her Daughter, Victoria, 1849–1855* (London: J. Cape, 1928), 43–44.

8

From the Civil War to the Space Age

The Civil War affected the lives and estates of many figures previously noted in this book. Edmund Ruffin, for instance, the editor of the *Farmer's Register,* was a staunch pro-slavery advocate. A strong supporter of the secession movement, he fired one of the first cannon shots of the Civil War at Fort Sumter in Charleston, South Carolina. When the South collapsed in 1865, he committed suicide. Fifty-seven-year-old John Cook Bennett, a zealous abolitionist, helped organize and recruit the Tenth Iowa Infantry Regiment but was invalided out after a few months because of a severe case of diarrhea, a disease that he professed could be cured by ingesting tomatoes.

The Landreths' seed store in Charleston was closed down by the Confederate government in 1861. Their grandson fought with distinction in the Union army. During the war, Martha Logan's letters, diaries, and other papers were sent from Charleston to Columbia, South Carolina, for safekeeping, but they were destroyed when General Sherman's troops torched the city in 1865. Southern truck farms that supplied northern cities before the war were devastated and did not recover for almost a decade. Tomato farmers in New Jersey, Delaware, Maryland, Pennsylvania, Indiana, and Ohio were stimulated by the demands of the Union armies. After the war, these states became major tomato growing and canning centers.

One of the easiest and quickest vegetables to grow, tomatoes were used as a food by both the Southern and Northern armies. Tomato recipes were published in cookbooks on both sides of the conflict. The effects of the war on tomato-related industries in the North were particularly dramatic. To feed the Northern army, contracts were let to canning factories, which employed mostly women. These contracts greatly stimulated the growth of tomato canneries. Confederate soldiers, who often endured with meager rations, fought not only for the Southern cause but also to acquire a square meal from a defeated Union army's supplies, which often included canned tomatoes. By the end of the war, empty tomato cans were found everywhere. While traveling west of the Mississippi River on his way to California in

1865, Samuel Bowles found canned tomatoes in "every hotel and station meal and at every private dinner or supper." They were used even in the mining camps in the Rocky Mountains in Colorado, where they sold for fifty cents to a dollar for a two-quart can.[1]

These wartime effects were small, however, compared to the dramatic expansion in tomato consumption after the war. Many Northern soldiers ate canned vegetables for the first time while they were in the army. After the war, the demand for canned products soared. In 1868 a single establishment in Philadelphia employed "four hundred women at a time for canning tomatoes and small fruits, and consumed in a single season $30,000 worth of sugar." Three factories in Burlington, New Jersey, employed six hundred people and disbursed several thousand dollars weekly for wages. These factories processed the production of nearly a thousand acres of tomatoes lying within a three-mile radius. By 1870 tomatoes were among the big three canned vegetables, along with peas and corn. By 1879 more than nineteen million cans of tomatoes were manufactured annually, generating revenue of over one million six hundred thousand dollars. Within a few years this production quadrupled. In subsequent years more tomatoes were canned than any other fruit or vegetable. By 1884 a single canning factory, J. H. Butterfoss in Hunterdon County, New Jersey, produced 340,000 cans of tomatoes and 43,000 gallons of catsup. In 1885 more than two million cases containing twenty-four cans each were produced nationally. Within three years more than five and a half million cases were produced. By the end of the century the number increased to almost fifteen million cases. In addition, the production of tomato ketchup expanded exponentially. By 1896 the *New York Tribune* proclaimed that tomato ketchup was the national condiment of the United States and was available on every table in the land.[2]

To meet the expanded need, farmers grew tomatoes in hothouses and used other techniques of forwarding them. By 1866 fresh tomatoes were available in many large cities throughout the year. With the completion of the transcontinental railroad, fresh tomatoes were shipped from California to New York in July 1869. As refrigerated train cars were not yet invented, it was perhaps no surprise that the tomatoes were a little overripe when they arrived. Railroads also helped California's agricultural competitors. Commercial production of tomatoes commenced in northern Florida in 1872. Within ten years Florida tomatoes were sold on the Chicago market for about four dollars and twenty-five cents per bushel crate. By 1889 they were grown as far south as the Florida Keys and were shipped northward weeks before local tomatoes ripened in northern states.[3]

During the Civil War, the sale of fresh tomatoes was so lucrative that farmers in Bermuda exported them to northern cities in May and the early summer months. In 1876 Bahamian farmers began to do the same. In these British colonies, tomatoes ripened several weeks before those growing in northern states, so the exporters were financially compensated for their shipping costs. By 1879 more than eight thousand boxes were exported from the Bahamas alone. Other islands in the Caribbean began to grow tomatoes and other vegetables for export into the United States.

To protect American growers against this competition, Congress passed the Tariff Act of 1883, levying a ten percent duty on imported vegetables. In the spring of 1886 John Nix imported tomatoes into New York from the West Indies. Maintaining that they were a fruit rather than a vegetable, Nix paid the duty under protest. In February 1887 he brought suit in New York against the collector, Mr. Hedden, to recover the duties. After six years of winding through courts and appeals courts, the case of *Nix v. Hedden* was argued before the Supreme Court. Nix's counsel read into evidence definitions of the words *fruit, vegetable,* and *tomato* from *Webster's Dictionary, Worcester's Dictionary,* and the *Imperial Dictionary.* He also called two witnesses who had been in the business of selling fruit and vegetables for thirty years and asked them, after hearing these definitions, to say whether these words had "any special meaning in trade or commerce, different from those read." One witness testified that the term *fruit* was "applied in trade only to such plants or parts of plants as contain the seeds." In May the opinion of the Supreme Court was delivered by Justice Horace Gray, who reported that the single question in this case was whether tomatoes, considered as provisions, were classed as *vegetables* or as *fruit* within the meaning of the Tariff Act of 1883. He concluded,

> Botanically speaking tomatoes are the fruit of a vine, just as are cucumbers, squashes, beans and peas. But in the common language of the people, whether sellers or consumers of provisions, all these are vegetables, which are grown in kitchen gardens, and which, whether eaten cooked or raw, are, like potatoes, carrots, parsnips, turnips, beets, cauliflower, cabbage, celery and lettuce, usually served at dinner in, with or after the soup, fish or meats which constitute the principal part of the repast, and not, like fruits generally, as dessert.

Based on this logic, the Supreme Court sided with the duty collector. Since then some pop historians have claimed gleefully but erroneously that the Supreme Court was unaware that the tomato was actually a fruit.[4]

The number of tomato varieties increased spectacularly. Alexander Livingston, who as a child was told by his mother that tomatoes were poisonous, became one of the most important developers of tomato varieties in America. By adhering to the principle of single-plant selection to clearly define demands in the tomato trade, he developed or improved thirteen major varieties from 1870 to 1893. Among the more important were the Paragon, Acme, Perfection, Golden Queen, Livingston's Favorite, and the Buckeye State Tomato. Many of Livingston's endeavors were conducted at Reynoldsburg, Ohio. In commemoration of his work, Reynoldsburg has claimed that it is the home of the commercial tomato. Since 1965 the town has held an annual tomato festival, which, of course, includes contests for the largest tomato plant, the heaviest fruit, the smallest fruit, and forty-one other categories.[5]

In 1886 Liberty Hyde Bailey, an agricultural researcher, located and tested seventy-six varieties sold by seedsmen. The following year he included in his tests a hundred and seventy sorts from American, British, French, and German seedsmen. This increase was due to several factors: the development of many new American varieties; the introduction of renamed European varieties; the tendency of seedsmen to list as distinct varieties those that differed little from already named ones; and the reluctance of seedsmen to remove duplicates from their list because of customer loyalty to particular names.[6]

Unfortunately, few of the tomato varieties cultivated in America before the Civil War have survived. The Oliver Kelly Farm in Elk River, administered by the Minnesota Historical Society, grows some tomato varieties that date to the 1850s. Heirloom seeds from the latter part of the nineteenth century are available today through several different seed companies and organizations. In Pennsylvania the Landis Valley Museum sells certain varieties of heirloom tomato seeds, and others are available through the Tomato Growers Supply Company in Fort Meyers, Florida, and the Southern Exposure Seed Exchange in North Garden, Virginia. In Iowa the Seed Savers Exchange was officially launched by Kent and Diane Whealy in 1975 expressly for the purpose of saving heirloom seeds through networking with growers all over the United States. Its *1993 Yearbook* lists more than a thousand varieties of tomato seeds.[7] The Federal government too has funded projects for storing tomato seeds. The Tomato Genetics Resource Center, led by Dr. Charles Rick, emeritus professor of vegetable crops at the University of California in Davis, has collected approximately a thousand tomato accessions from wild species in South America. The center has a total of 2,400 entities or accessions. The National Seed Storage Laboratory in Geneva, New

York, houses more than 5,500 tomato accessions. Despite these efforts, large numbers of tomato varieties known to have existed during the nineteenth century have simply disappeared.

So has much of the tomato cookery propagated during that period. Tomato cookery in America was in its heyday throughout the late nineteenth and early twentieth century. By 1893 A. W. Livingston had published more than sixty-five tomato recipes. Fannie Merritt Farmer's *Boston Cooking-school Cook Book,* first published in 1896, included scores of recipes with tomatoes as ingredients. The creativity reflected in early tomato recipes waned during the early part of the twentieth century as attempts to breed tomatoes that were easy to transport and had a long shelf life led to the creation of the unpalatable tomato varieties whose progeny haunt our supermarkets today.[8] While these trends made the tomato more accessible in an America, which was rapidly moving from rural to urban areas, taste and variety were sacrificed. Obviously these problems were not limited to the tomato but reflected general trends affecting most fruits and vegetables.

Tomatoes have not shed their controversial character. After the Civil War, Dio Lewis continued to question the tomato's purported healthful qualities. In one lecture he asked, "Didn't you know that eating bright red tomatoes caused cancer?" Pennsylvania physician Dr. John Hylton took this point up, claiming that tomato cells were identical to cancer cells under the microscope. He further asserted that "there was much cancer where tomatoes were eaten." Dr. Homer Hitchcock sent out an inquiry to the physicians in Michigan asking for their evaluation of Hylton's claims. All responded negatively. One physician, Dr. John Folly, stated that "tomatoes will make cancer as soon as a red thread around the neck will stop the nosebleed." Despite the fact that this charge was denied by many physicians and other prominent individuals, it lingered for years. As late as the mid-1930s Della Lutes reported that cancer "was a word with which we were somewhat familiar, since some of our more conservative neighbors would not touch tomatoes on account of how Ob Hutchens, who ate them, had developed this dread disease."[9]

Other physicians attacked the proclaimed virtues of the tomato during the late nineteenth century. As late as 1896 the American Medical Association debated its healthfulness. W. T. English, a professor of physical diagnosis in the medical department at Western University of Pennsylvania, declared that ingesting tomatoes caused high blood pressure, sighing respiration, skipped heartbeats, cold sweats, and disturbed vision. English maintained that general depression was "a common sequence to their employment in one of the alms houses near Pittsburgh" and in several hospitals. "The supplies for the military should never include this vegetable," he declared, for "it can not be for the physical good of the men." The tomato was "treacherous" and could "not be relied upon as a food." Those called upon to endure the loss of sleep or suffer mental strain, he said, such as "the student, the speaker, the salesman, will find them worse than useless."[10]

The healthful qualities of the tomato were not established until the mid-twentieth century. As has been previously noted, the tomato does have considerable vitamin C and some vitamin A. In a survey conducted by M. Allen Stevens of the University of California at Davis, the tomato ranked thirteenth among other commonly consumed fruits and vegetables as a source of vitamin C and sixteenth as a source of vitamin A. Overall it ranked sixteenth nutritionally behind such vegetables as broccoli, spinach, brussels sprouts, lima beans, peas, asparagus, artichokes, cauliflower, sweet potatoes, carrots, sweet corn, potatoes, and cabbage. Paradoxically, despite this comparatively low ranking, tomatoes contributed more vitamin A and C and other nutrients to the American diet than did other fruits and vegetables because so many more tomatoes were consumed. Today, medical researchers believe that tomatoes may be a potential anti-cancer weapon. Cancer researcher Frank Meyskens of the School of Medicine at the University of California at Irvine recently reported that tomatoes contain lycopene, a carotene that gives tomatoes their red color. New research has demonstrated that people with more lycopene in their blood have lower rates of certain cancers. Tomatoes are almost the only source of lycopene in our diets. [11]

During the latter part of the twentieth century, tomato controversies have taken new twists. Spoofs on the old belief about the dangers of tomatoes, the cult film *The Attack of the Killer Tomatoes* and its sequel have spawned a popular Frankensteinesque cartoon show of the same name. Killer tomato paraphernalia can be acquired in toy stores throughout America. A real controversy was related to the "space tomatoes." In 1984 the National Aeronautics and Space Administration sent more than twelve million tomato seeds into space, where they remained for six years. The space shuttle *Columbia* retrieved the seeds, and NASA distributed them to teachers to grow in their classrooms. Public uproar followed. Parents, teachers, and scientists expressed fears about the potential mutations of the space tomatoes. Gregory Marlins, the director of the NASA seed project, recommended that students not eat the tomatoes because of the possibility that some potential mutations might be toxic.

A much more serious controversy erupted with regard to the genetic engineering of new tomato varieties. Tomato genes have been mapped extensively during the past thirty years. Each of the tomato's twelve chromosomes can be easily recognized during certain stages of its reproduction. One noteworthy genetic trait of the tomato is that generally each of its functions is controlled by only one gene. This combination of characteristics has made the tomato a relatively easy organism in which to locate exact chromosomal positions, a condition that greatly aids engineering techniques. Bioengineers maintain that through selective breeding and naturally occurring mutations they have produced genetically engineered tomatoes with some of the qualities that have been sought for years.

Beginning in the 1950s, botanists induced genetic mutations with X-rays and chemicals. These mutants were mainly of interest to researchers. The research, however, encouraged further investigations into the chromosomal structure of the tomato, making more sophisticated alterations possible. During the past few years this research has become productive. In a project funded by Campbell's Soup Company, Calgene, Inc. in Davis, California, genetically engineered the first tomato, called MacGregor's. Calgene claims that MacGregor's is slow-ripening, "terrific tasting," and transportable over great distances without loss of quality. The variety has received a patent, and the Food and Drug Administration has permitted it to be sold to the public. Other genetically engineered tomatoes, such as the Flavor Savr, are under development. Another bioengineering firm, DNA Plant Technology of Cinnaminson, New Jersey, has "synthesized genetic material from fish that showed promise in protecting tomatoes from freezing." Companies such as Petroseed, Monsanto, Pioneer, and DuPont are not far behind.[12] Many people, including the futurist Jeremy Rifkin, have strongly opposed genetic engineering of the tomato as presenting an unacceptable risk for humans. Some grocery stores have refused to sell genetically engineered tomatoes, while others have agreed to identify them as genetically engineered. More than two thousand five hundred restaurants have announced that they will boycott the new "mutant" tomato. These controversies will not go away soon.

As we approach the twenty-first century, though, the news is not all bad or controversial. More Americans are growing tomatoes in their home gardens than ever before. Modern tomato varieties are available that can be grown in most climatic and soil conditions. Heirloom seeds are readily acquirable for those interested in reexperiencing that old-fashioned tomato flavor or tasting it for the first time.

Though by and large the interest in creative tomato cookery waned in the early part of the twentieth century, there were and continue to be resurgences of inspiration. Southern Italian immigration during the late nineteenth and early twentieth century brought new ways of using the tomato to America. Recent Hispanic immigrations from the Caribbean, Mexico, and Central and South America have had a similar influence on tomato cookery in the late twentieth century. One of the fastest-growing food products today is tomato-based salsa, which in 1991 surpassed tomato ketchup in sales. Thanks in part to these influences, tomato cookery has renewed vigor. Tomato cookbooks have been published recently, and more are in development. As migrations from other tomato-eating cultures are likely to increase, the Cinderella-like, rags-to-riches story of the tomato in America has not yet ended.

NOTES

1. *The Volunteer's Cook-book: for the Camp and March* (Columbus, Ohio: Joseph H. Riley, 1861), 31; Judy and Hugh Gowan, *Blue and Grey Cookery: Authentic Recipes from The Civil War Years* (New Market, Md.: Daisy Publications, n.d.), 4, 26; *Confederate Receipt Book; A Compilation of over One Hundred Receipts, Adapted to the Times,* reprint (Athens: University of Georgia Press, 1960), 18; Mrs. Annie Wittenmeyer, *A Collection of Recipes for the Use of Special Diet Kitchens in Military Hospitals,* reprint, (Mattituck, N.Y.: J. M. Carroll, 1983), 11–12, 28; Carlton McCarthy, "Detailed Minutiae of a Soldier's Life," *Southern Historical Society Papers,* 6 (July 1878): 5; Samuel Bowles, *Across the Continent* (Springfield, Mass.: Samuel Bowles, 1866), 64–65; Bowles, *Our New West: Records of Travel* (Hartford, Conn.: Hartford Publishing, 1869), 199–200.

2. *Country Gentleman,* 32 (December 3, 1868): 376; *Horticulturist,* 23 (November 1868): 321–25; *Maryland Farmer,* 16 (August 1879): 261; "Report of the Hunterdon County Agricultural Society," as in the *Country Gentleman,* 49 (August 21, 1884): 696; *Historical Statistics of the United States; Colonial Times to 1970* (Washington, D.C.: U.S. Department of Commerce; Bureau of the Census, 1975), vol. 2, 690–91; *New York Tribune,* as in the *Scientific American Supplement,* 42 (November 26, 1896): 17435.

3. *American Agriculturist,* 28 (August 1869): 283; George F. Weber, "A Brief History of Tomato Production in Florida," *Proceedings of the Florida Academy of Sciences,* 4 (1940): 167–74.

4. Paul Albury, *The Story of the Bahamas* (Hong Kong: Macmillan Education, 1975), 187–88; United States Reports, *Cases Adjudged in The Supreme Court at October Term, 1892* (New York: Banks Law Publishing, 1910), vol. 149, 304–8.

5. Alexander W. Livingston, *Livingston and the Tomato* (Columbus, Ohio: A. W. Livingston's Sons, 1893); *Tomato Facts: The Story of the Evolution of the Tomato* (Columbus, Ohio: Livingston Seed, 1909).

6. Liberty Hyde Bailey, "Notes on Tomatoes," Agricultural College of Michigan *Bulletin* No. 19 (Lansing: Thorp & Godfrey, 1886); Liberty Hyde Bailey, "Notes on Tomatoes," Agricultural College of Michigan *Bulletin* No. 31 (Lansing: Thorp & Godfrey, 1887), 13–25; Gordon Morrison, "Tomato Varieties," Agricultural Experiment Station Special *Bulletin* No. 290 (East Lansing: Michigan State College, April 1838), 1–7.

7. *Seed Savers 1993 Yearbook* (Decorah, Iowa: Seed Savers Exchange, 1993), 251–306.

8. Livingston, *Livingston and the Tomato,* 138–57; Fannie Merritt Farmer, *Boston Cooking-school Cook Book*, reprint (New York: Weathervane Books, 1986).

9. Will W. Tracy, *The Tomato Culture* (New York, Orange Judd, 1907), 7; *Eclectic Medical Journal of Pennsylvania,* 9 (September–October 1871): 286; Homer O. Hitchcock, "Cancer not Caused by Tomatoes," *Annual Report of Michigan State Board of Health,* 6 (1878): 35–38; *Country Gentleman,* 53 (August 9, 1888): 600; W. C. Latta, *Outline History of Indiana Agriculture* (Lafayette, Ind.: Alpha Lambda Chapter of Epsilon Sigma Phi, 1938), 264; Della Lutes, *The Country Kitchen* (Boston: Little, Brown, 1936), 21.

10. *Journal of the American Medical Association,* 26 (January–June 1896): 1255–58.

11. Jean Carper, "Tomatoes Gain New Respect as an Anti-Cancer Food," *Tampa Tribune-Times,* March 15, 1992; Charles Rick, "The Tomato," *Scientific American,* 239 (August 1978): 78.

12. Charles Rick, "Genetic Resources in Lycopersicon," in Donald J. Nevins and Richard A. Jones, *Tomato Biotechnology* (New York: Alan R. Liss, 1987), 19; Benedict Carey, "Tasty Tomatoes: Now There's a Concept," *Health,* 7 (July-August 1993): 26; Donald Woutat, "Toward a Tastier Tomato," *Los Angeles Times,* July 7, 1993; John Seabrook, "Tremors in the Hot House," *New Yorker,* 69 (July 19, 1993): 32–41.

Part II

Historical Recipes

Painting by Paul Lacroix, dated 1863. *(The location of the original painting is unknown; photograph in the possession of Dr. William Gerdts.)*

Historical Recipes

Thousands of recipes including tomatoes among their ingredients were published or written in the United States prior to the Civil War. They appeared in cookbooks, agricultural and horticultural journals, newspapers, almanacs, seed catalogues, medical works, magazines, and a host of other sources. In this section there is a representative sample of those recipes. Some were selected because they were typical, others because they were unusual. English, French, Italian, and even Turkish recipes are included because they were commonly published in America. Some recipes were experiments that did not survive into the twentieth century, while others survive today. As a collection they reflect the diversity of the tomato and demonstrate the variety of sources in which recipes appeared. These recipes differ markedly from those that appear in modern cookbooks. Specific cooking times were not always given. Because of the state of cooking technology in the early nineteenth century, cooks were unable to control temperatures easily, and thus it was difficult to specify exact cooking times. Quantities often depended upon what was available. As tomatoes were not uniform, quantities depended too upon their size, shape, consistency, and such other factors as their acid content and their taste and color. When quantities were listed, they were often very large. Proportionally scaling back the quantities in these recipes will often result in finished products that taste similar to the originals.

Throughout the nineteenth century, cookery was truly an art. Extensive experience was needed to perform even basic cooking functions. Most cookbook authors assumed that readers already possessed this experience and that there was therefore little need to spell everything out. The cook was expected to do what made sense rather than blindly follow the directions of a cookbook author who had no idea of specific cooking conditions, equipment, or the availability of ingredients in the case of any given cook.

Spelling and directions in these recipes have been left in their original form. There are a few measures and terms in them that are not commonly used today. For instance, a *gill* is a liquid measure equal to a quarter of a pint. A *peck* is equal to eight quarts. A *bushel* is equal to four pecks, or thirty-two quarts. A *pottle* is equal to half a gallon. A *cimlin,* or *cymling,* is a pattypan squash. A *salamander* is a utensil made broiling hot in a fire for browning pastry. A *tammy* (*tamy, tammis,* or *tamis*) is a cloth used for straining liquids.

SALAD, SOUP, GAZPACHO, AND GUMBO

SALAD

From the middle of the sixteenth century, herbalists frequently mentioned that tomatoes could be eaten like cucumbers—sliced up and dressed with oil and spices. Throughout the early nineteenth century, accounts of eating raw tomatoes appeared regularly in a variety of American sources. Some writers recommended seasoning them with sugar, molasses, vinegar, salt, pepper, mustard, or milk. Others ate them without any seasoning at all. Lettice Bryan's recipe recommended eating raw tomatoes for breakfast or dinner. Today's typical lettuce and tomato salad is a product of the late nineteenth century.

To Dress Tomatoes Raw—Lettice Bryan, 1841

Take ripe tomatoes, that are large and fine, peel and slice them tolerably thick, put them in a deep dish, and season them highly with salt, pepper and vinegar. This is a delicious breakfast dish, and is also a fine accompaniment to roast meats, for a dinner.

From Lettice Bryan, *The Kentucky Housewife.* Cincinnati: Stereotyped by Shepard & Stearns, 1841. 217.

SOUPS

In England and Colonial America, tomatoes were used as ingredients in soup at least as early as the mid-eighteenth century. The first known American recipe titled "Tomato Soup" was published by N. K. M. Lee. It was really a vegetable soup containing carrots, celery, onions, and turnips. Lettice Bryan's recipe is closer to what we think of as tomato soup.

Tomata Soup—N. K. M. Lee, 1832

Wash, scrape, and cut small the red part of three large carrots, three heads of celery, four large onions, and two large turnips, put them into a saucepan, with a table-spoonful of butter, and half a pound of lean new ham; let them stew very gently for an hour, then add three quarts of brown gravy soup, and some whole black pepper, with eight or ten ripe tomatas; let it boil an hour and a half, and pulp it through a sieve; serve it with fried bread cut in dice.

From N. K. M. Lee, *The Cook's Own Book.* Boston: Munroe and Francis, 1832. 222.

Tomato Soup—Lettice Bryan, 1841

Peel and slice two quarts of tomatoes, boil them till dissolved in two quarts of beef, veal or poultry broth; then strain it into a soup-pan, put in a few more tomatoes that have been peeled and sliced, add a large lump of butter, and pepper and salt, to your taste; boil it till the tomatoes are done, stirring it frequently, and serve with it a plate of dry toasts or crackers.

From Lettice Bryan, *The Kentucky Housewife.* Cincinnati: Stereotyped by Shepard & Stearns, 1841. 24–25.

GAZPACHO

During the first half of the nineteenth century, many soup recipes were imported from other nations or cultures. Gazpacho is a traditional Spanish dish, but the first known gazpacho recipe may have been published in America. Two are included here, one from Mary Randolph and the other from Louis Eustache Audot.

Gaspacha—Spanish—Mary Randolph, 1824

Put some soft biscuit or toasted bread in the bottom of a salad bowl, put in a layer of sliced tomatas with the skin taken off, and one of sliced cucumbers, sprinkled with pepper, salt, and chopped onions; do this until the bowl is full, stew some tomatas quite soft, strain the juice, mix in some mustard and oil, and pour over it; make it two hours before it is eaten.

From Mary Randolph, *The Virginia House-wife.* Washington: Davis and Force, 1824. 107.

Gaspacho—Louis Eustache Audot, 1846

Take two onions, some tomatas, a handful of green pimento, a cucumber, a clove of garlic, parsley and chervil; cut the whole into small pieces, and put them into a salad-bowl. Add as much crumbed bread as will form double the quantity which the dish already contains; season with salt, pepper, oil, and vinegar, like a salad, and complete the *gaspacho* with a pint of water to make the *bouillon. Gaspacho* is eaten with a spoon; it is a kind of raw soup. It is a favourite dish with the Andalusians, and is very refreshing and wholesome in their warm climate.

From Louis Eustache Audot, *French Domestic Cookery.* New York: Harper & Brothers, 1846. 272–73.

Gumbo

The okra plant originated in West Africa and may have been brought to America by slaves from the Caribbean. Some recipes for gumbo, an okra-based Creole soup or stew, incorporated tomatoes as an ingredient. Mrs. Read's recipe for tomato and okra soup was similar to many that circulated in America and were most likely influenced by French Creole cookery from the Caribbean or New Orleans. Mary Randolph's recipe for okra soup includes six tomatoes and cimlins, or pattypan squash. In Randolph's recipe, the okra is stewed with skinned tomatoes. Eliza Leslie's recipe for beef gumbo is typical of the gumbo recipes containing tomatoes.

To Make Tomatoe & Ochre Soup—Mrs. Read, 1813

One pound of Beef—1 doz Ochre—1 doz Tomatoes—one Onion, half a green pepper—early in the morning, put the Beef & Ochre in 3 gallons of water—and boil it to 3 pints—Skim it well add the Tomatoes &c—about 1 Oclock, for Diner at 3.

From manuscript cookbook, Mrs. George Read, New Castle, 1813, Holcomb Collection, Historical Society of Delaware, Wilmington.

Ochra Soup—Mary Randolph, 1824

Get two double handsful of young ochra, wash and slice it thin, add two onions chopped fine, put it into a gallon of water at a very early hour in an earthen pipkin, or very nice iron pot: it must be kept steadily simmering, but not boiling: put in pepper and salt. At 12 o'clock, put in a handful of Lima beans, at half past one o'clock, add three young cimlins cleaned and cut in small pieces, a fowl, or knuckle of veal, a bit of bacon or pork that has been boiled, and six tomatas, with the skin taken off when nearly done; thicken with a spoonful of butter, mixed with one of flour. Have rice boiled to eat with it.

From Mary Randolph, *The Virginia House-wife.* Washington: Davis and Force, 1824. 34–35.

Ocra and Tomatas—Mary Randolph, 1824

Take an equal quantity of each, let the ocra be young, slice it, and skin the tomatas, put them into a pan without water, add a lump of butter, an onion chopped fine, some pepper and salt, and stew them one hour.

From Mary Randolph, *The Virginia House-wife.* Washington: Davis and Force, 1824. 95–96.

Beef Gumbo—Eliza Leslie, 1850

Put into a large stew-pan some pieces of the lean of fresh beef, cut up into small bits, and seasoned with a little pepper and salt. Add sliced ochras and tomatas, (either fresh, or dried ochras and tomata paste.) You may put in some sliced onions. Pour on water enough to cover it well. Let it boil slowly, (skimming it well,) till every thing is reduced to rags. Then strain and press it through a cullender. Have ready a sufficiency of toasted bread, cut into dice. Lay it in the bottom of a tureen, and pour the strained gumbo upon it.

From Eliza Leslie, *Miss Leslie's Lady's New Receipt-book.* Philadelphia: A. Hart, Late Carey & Hart, 1850. 445–46.

MAIN COURSES

BAKED, SCOLLOPED, FRIED, AND STEWED TOMATOES

Sarah Rutledge's recipe typifies baked tomato recipes prior to the Civil War. Her method was similar to scolloping tomatoes, or baking them in layers with bread crumbs, butter, and seasonings. Mary Randolph's recipe "To Scollop Tomatas" was frequently borrowed by other cookbook writers. Baked vegetables, particularly tomatoes, became a traditional dish on the southern table during the latter part of the nineteenth century. There were many ways of frying and stewing tomatoes. Lettice Bryan's recipe is similar to ways that tomatoes are fried today. Fanny Flagg's book *Fried Green Tomatoes at the Whistle Stop Cafe* and the subsequent movie rejuvenated interest in this dish, which probably had its origin in the northern states. Mary Randolph's recipe for stewed tomatoes is very simple. Others included a variety of seasonings. Throughout the 1840s and 1850s stewed tomatoes were served in restaurants, either alone or combined with other foods.

To Scollop Tomatas—Mary Randolph, 1825

Peel off the skin from large, full, ripe tomatas—put a layer in the bottom of a deep dish, cover it well with bread grated fine; sprinkle on pepper and salt, and lay some bits of butter over them—put another layer of each, till the dish is full—let the top be covered with crumbs and butter—bake it a nice brown.

From Mary Randolph, *The Virginia House-wife.* 2d ed. Washington: Way & Gideon, 1825. 140.

To Bake Tomatoes—Sarah Rutledge, 1847

Scald and peel about a dozen or more ripe tomatoes; butter a shallow baking-dish, and put in the finest without breaking them, and not quite touching; fill up the little space between with small pieces of stale bread, buttered. The rest of the tomatoes mash, and strain out all the hard parts; then mix with a spoonful of butter, pepper and salt. Pour it over the dish, strew bread-crumbs on the top. Bake about half an hour.

From Sarah Rutledge, *The Carolina Housewife.* Charleston: W. R. Babcock, 1847. 103.

Fried Tomatoes—Lettice Bryan, 1841

Select them large and ripe, take off the peelings, cut them in thick slices, and season them with salt and pepper. Have ready a plate of finely grated bread, dip each side of the sliced tomatoes in it, taking care to make as much of the bread adhere to them as possible, and fry them brown in butter, which should be hot when they are put in. Serve them warm; mince very fine an onion or two, fry them in the gravy, and transfuse the whole over the tomatoes.

From Lettice Bryan, *The Kentucky Housewife.* Cincinnati: Stereotyped by Shepard & Stearns, 1841. 217.

To Stew Tomatas—Mary Randolph, 1825

Take off the skin, and put them in a pan with salt, pepper, and a large piece of butter—stew them till sufficiently dry.

From Mary Randolph, *The Virginia House-wife.* 2d ed. Washington: Way & Gideon, 1825. 141.

EGGS

Eggs and tomatoes were a favorite combination in Spain and the Caribbean by the mid-eighteenth century. Mary Randolph's recipe was among the first published in America that combined these ingredients. It was also one of the simplest. In the *American Agriculturist*'s recipe for a tomato omelet, the tomato and egg combination is browned on one side.

Eggs and Tomatas—Mary Randolph, 1824

Peel the skins from a dozen large tomatas, put four ounces butter in a frying pan, add some salt, pepper and a little chopped onion, fry them a few minutes, add the tomatas and chop them while frying; when nearly done, break in six eggs, stir them quickly, and serve them up.

From Mary Randolph, *The Virginia House-wife.* Washington: Davis and Force, 1824. 107.

*To make Tomato Omelet—*American Agriculturist, *1846*

Take a stew-pan and melt a piece of butter the size of nutmeg. Mince up an onion very fine, and fry it until quite brown. Add ten peeled tomatos, season with pepper and salt, and stir them until cooked to a soft pulp. Then stir in four beaten eggs, until the underside of the mass becomes brown. Lay a plate on top, turn the pan upside down, and the dish is ready for the table.

From *American Agriculturist,* 5 (September 1846): 269.

Fish and Seafood

Tomatoes were a common ingredient in fish and seafood recipes. The first such recipe published in America was in a cookbook by Richard Briggs. Sarah Rutledge's recipe for "Baked Shrimps and Tomatoes" calls for alternating layers of shrimp and stewed tomatoes. Pickled tomatoes are an ingredient in Eliza Leslie's "Sea Bass with Tomatoes."

To Dress Haddock the Spanish Way—Richard Briggs, 1792

Take two fine haddocks, scale, gut, and wash them well, wipe them with a cloth, and broil them; put a pint of sweet oil in a stew-pan, season it with pepper and salt, a little cloves, mace, and nutmeg beaten two cloves of garlick chopped, pare half a dozen love-apples and quarter them, when in season, put them in, and a spoonful of vinegar, put in the fish, and stew them very gently for half an hour over a slow fire; put them in a hot dish, and garnish with lemon.

From Richard Briggs, *The New Art of Cookery.* Philadelphia: W. Spotswood, R. Campbell, and B. Johnson, 1792. 80.

Baked Shrimps and Tomatoes—Sarah Rutledge, 1847

Butter well a deep dish, upon which place a thick layer of pounded biscuit. Having picked and boiled your shrimps, put them upon the biscuit; a layer of shrimps, with small pieces of butter, a little pepper, mace or nutmeg. On the top of the shrimps put a layer of stewed tomatoes, with a little butter, pepper and salt. Then add a thinner layer of beat biscuit, and another of shrimps, and so on, till three or four layers of both are put in the dish. The last layer must be of biscuit. Bake, and brown the whole.

From Sarah Rutledge, *The Carolina Housewife.* Charleston: W. R. Babcock, 1847. 55–56.

Sea Bass with Tomatoes—Eliza Leslie, 1847

Take three large fine sea-bass, or black-fish. Cut off their heads and tails, and fry the fish in plenty of lard till about half done. Have ready a pint of tomatoes, that have been pickled cold in vinegar flavoured with a muslin bag of mixed spices. Drain the tomatoes well from the vinegar; skin them, and mash them in a pan; dredging them with about as much flour as would fill a large table-spoon heaped up. Pour the mixture over the fish while in the frying pan; and continue frying till they are thoroughly done.

Cutlets of halibut may be fried in this manner with tomatoes: also, any other pan-fish.

Beef-steaks or lamb-chops are excellent fried thus with tomatoes.

From Eliza Leslie, *The Lady's Receipt-book.* Philadelphia: Carey and Hart, 1847. 22–23.

Poultry, Meat, and Sweetbreads

Tomatoes were cooked with all types of poultry and meat. They were often added as a condiment after cooking, as in N. K. M. Lee's recipe. Eliza Leslie's recipe for chicken and tomatoes demonstrates the Spanish influence in American tomato cookery. Her "Tomato Sweetbreads" is a common example of a main course frequently served in America's finest restaurants during the 1840s.

Chickens and Tomata Sauce—N. K. M. Lee, 1832

Mix together, in a stewpan, a little butter, salt, pepper, lemon-juice, and grated nutmeg, a sufficient quantity to put in two chickens; tie it in, and lay thin slices of lemon on the breast of the chickens, and lay them in a stewpan lined with thin rashers of bacon; cover them with the same, and stew them with fire above and below for three quarters of an hour; when done, drain them in a cloth; untie them, and serve with tomata sauce.

From N. K. M. Lee, *The Cook's Own Book and Housekeeper's Register.* Boston: Munroe & Francis, 1832. 50.

Pollo Valenciano—Eliza Leslie, 1850

This is also a Spanish dish. Cut up a large fine fowl into pieces. Wipe them clean and dry, but do not wash them or lay them in water. Put into a broad sauce-pan, a tea-cup of sweet oil, and a bit of bread. Let it fry, (stirring it about with a wooden spoon,) and when the bread is browned take it out, and throw it away. Then put in a sliced onion, and fry that; but take care not to let it burn or it will become bitter, and spoil the stew. Then put in the pieces of fowl, and let them brown for a quarter of an hour. Then transfer it to a stew-pan, adding a little bit of chili or red pepper minced small, and some chopped sweet herbs. Also half a dozen large tomatas quartered; and two tea-cups of *boiled* rice. Add a little salt, and stir the whole well together, having poured on sufficient hot broth to cover it. Place it over the fire, and when it has come to a hard boil, put the lid on the pan and set it aside to simmer till the whole is completely cooked, and the gravy very thick. About ten minutes before the stew goes to table, take off the lid of the stew-pan, lest the steam should condense on it and clod the rice, or render it watery. Serve it up *un*covered.

From Eliza Leslie, *Miss Leslie's Lady's New Receipt-book.* Philadelphia: A. Hart, Late Carey & Hart, 1850. 440–41.

Tomato Sweetbreads—Eliza Leslie, 1847

Cut up a quarter of a peck (or more) of fine ripe tomatoes; set them over the fire, and let them stew with nothing but their own juice till they go entirely to pieces. Then press them through a sieve, to clear the liquid from the seeds and skins. Have ready four or five sweetbreads that have been trimmed nicely, cleared from the gristle, and laid open to soak in warm water. Put them into a stew-pan with the tomato-juice, seasoned with a little salt and cayenne. Add two or three table-spoonfuls of butter rolled in flour. Set the sauce-pan over the fire, and stew the sweetbreads in the tomato-juice till they are thoroughly done. A few minutes before you take them off, stir in two beaten yolks of eggs. Serve up the sweetbreads in a deep dish, with the tomato poured over them.

From Eliza Leslie, *The Lady's Receipt-book.* Philadelphia: Carey and Hart, 1847. 66–67.

DUMPLINGS

The *American Farmer*'s recipe for tomato dumplings is unusual in that it was one of the few recipes to originate in an agricultural periodical and later be published in cookbooks. Even more surprising, the first cookbook to include it was written by the British author Eliza Acton. While there were many recipes in which tomatoes were filled with almost every conceivable ingredient, this was one of the few recipes in which tomatoes were placed in dough envelopes.

Tomatoes Dumpling—American Farmer, *1842*

Although we have but little faith in the belief that the inventor of the tomatoes pills will ever be able to substitute his concentrated extract of tomato for calomel, yet we verily do most conscientiously believe, that the day is not distant, when tomatoes dumplings and puddings will be just as *fashionable* on the dinner table, as *bustles* now are with the ladies. In the manner of composition, mode of cooking, and sauce, the good housewife must proceed the same as she would with an apple dumpling, with this exception, that care must be taken in *paring* the tomato, not to extract the seed, or break the *meat*, in the operation of skinning it. We have eaten this vegetable raw, without anything—cut up with vinegar, salt, pepper and mustard—fried in butter, and in lard, broiled and basted with butter, stewed with, and without bread, with cream and with butter—and with a clear conscience we can say, we like them in every way they have been ever fixed for our palate; but of all the modes of dressing them known to us, we prefer them when cooked in dumplings, for to us it appears that the *steaming* they receive in their dough envelopes, increases in a very high degree that delicate spicy flavor which, even in their uncooked state, make them such decided favorites of the epicure.

From *American Farmer,* 3d ser., 4 (August 17, 1842): 101.

STUFFED AND FORCED TOMATOES

Tomatoes were stuffed with almost all types of available ingredients. Eliza Leslie stuffed them with grated bread, herbs, and seasonings. *The Lady's Annual Register*'s recipe, translated directly from the French, called for filling the tomatoes with thyme, garlic, and other ingredients. Eliza Acton's French recipe included meats, egg yolks, and mushrooms. Charles Elmé Francatelli's recipe, "Tomatas a la Provencale," involved a more complex process of preparation.

Stuffed Tomatas—Eliza Leslie, 1832

Scoop out the inside of a dozen large tomatas, without spoiling their shape. Pass the inside through a sieve, and then mix it with grated bread, chopped sweet-herbs, nutmeg, salt, and pepper. Stew it ten minutes, with a laurel leaf, or two peach leaves. Remove the leaves, and stuff the tomatas with the mixture, tying a string round each to keep them in shape. Sprinkle them all over with rasped bread-crust. Set them in a buttered dish, and bake them in an oven. Take off the strings, and serve up the tomatas.

From Eliza Leslie, *Domestic French Cookery, Chiefly Translated from Sulpice Barué.* Philadelphia: Carey & Hart, 1832. 73–74.

Tomatos Stuffed—The Lady's Annual Register, *1843*

Take twenty four Tomatos very ripe and sound. Take off the stems, cut them in two, take out the seeds, lay them, the flower side downward, upon a tin sheet, without their touching each other. The side on which is the stem should be pressed hard into a saucepan with a piece of butter, two slices of ham, season with salt, pepper, a laurel leaf, a little thyme and garlic. Put the saucepan over a moderate fire, cook the Tomatos till they are of the consistency of soup; while they are stewing add a glass of wine and some leaves of parsley. When it is of a proper consistency, pass them through a hair sieve with the hand, scrape the bottom of the sieve with the back of a knife. Put this mixture into a saucepan with some crumbs of grated bread, Parmesan cheese, and a little coarse pepper; add to it a small quantity of olive oil, and fill the tomatos with it, and shake over them bread crumbs and Parmesan cheese. Moisten them well with oil, bake them in a hot oven. When done, lay them on a plate and serve them. (This receipt, which is a literal translation from one in a French cookery book, is not perfectly clear, but may give some hints to those accustomed to using Tomatos.)

From *The Lady's Annual Register, and Housewife's Almanac, for 1843.* Boston: T. H. Carter, 1843. 59.

Forced Tomatas (French Receipt)—Eliza Acton, 1845

Let the tomatas be well shaped and of equal size; divide them nearly in the middle, leaving the blossom-side the largest, as this only is to be used; empty them carefully of their seeds and juice, and fill them with the following ingredients, which must previously be stewed tender in butter, but without being allowed to brown; minced mushrooms and shalots, with a moderate proportion of parsley, some lean of ham chopped small, a seasoning of cayenne, and a little fine salt, if needed; let them cool, then mix with them about a third as much of fine crumbs of bread, and two yolks of eggs; fill the tomatas, cover them with fine crumbs, moisten them with clarified butter, and bake them in a brisk oven until they are well coloured. Serve them as a garnish to stewed rump or sirloin of beef, or to a boned and forced leg mutton.

Minced lean of ham, 2 ozs.; mushrooms, 2 ozs.; bread-crumbs, 2 ozs.; shalots, 4 to 8; parsley, full teaspoonful; cayenne, quarter saltspoonful; little salt, if needed; butter, 2 ozs; yolks of eggs 2 to 3; baked, 10 to 20 minutes.

Obs—The French pound the whole of these ingredients with a bit of garlic, before they fill the tomatas with them, but this is absolutely not necessary, and the garlic, if added at all, should be parboiled first, as its strong flavour, combined with that of the eschalots, would scarcely suit the general taste. When the lean of a dressed ham is at hand, only the herbs and vegetables will need to be stewed in the butter; this should be mixed with them into the forcemeat, which an intelligent cook will vary in many ways.

From Eliza Acton, *Modern Cookery in All its Branches . . . Prepared for American Housekeepers. By Mrs. S. J. Hale.* From the 2d London edition. Philadelphia: Lea and Blanchard, 1845. 241–42.

Tomatas, a la Provencale—Charles Elmé Francatelli, 1846

Slice off that part of the tomata that adheres to the stalk, scoop out the seeds without breaking the sides of the fruit, and place this in circular order in a saucepan, containing about a gill of salad oil. Next, chop up half a pottle of mushrooms, a handful of parsley, and four shalots; put these into a stewpan with two ounces of scraped fat bacon, and an equal portion of lean ham, either chopped or grated fine; season with pepper and salt, and a little chopped thyme. Fry these over the stove-fire for about five minutes; then, mix in the yolks of four eggs, fill the tomatas with this preparation, shake some light-colored raspings of bread over them, and place them over a brisk stove-fire, holding a red-hot salamander over them for about ten minutes, by which time they will be done; dish them in the form of a dome, pour some brown Italian sauce (No. 12) round the base, and serve.

From Charles Elmé Francatelli, *French Cookery. The Modern Cook.* Philadelphia: Lea and Blanchard, 1846. 372.

SAUCES, KETCHUPS, AND OTHER CONDIMENTS

TOMATO HODGE-PODGE

A hodge-podge was a dish made of a mixture of various kinds of meats or vegetables. *Hodge-podge* had been used as a culinary term since the early seventeenth century. Ann Allen's recipe was one of the few published prior to the Civil War that included tomatoes. It produces a basic green tomato relish that is used in ways that ketchups and sauces are used.

Tomato Hodge-Podge—Ann H. Allen, 1845

2 qts. of green tomatoes,
2 qts. of green peppers,
2 qts. of onions,
1 cup of salt,
1 pint of mustard-seed.

Cut all up fine, mix all well together, cut like mince-meat, then have a nice jar, and cover two inches thick, then strew salt and mustard-seed, then mince until through, set it away, and let it stand until it works a trifle, then put one quart of the best of vinegar over. It is excellent with meats.

From Ann H. Allen, *The Housekeepers' Assistant, Composed upon Temperance Principles*. Boston: James Munroe, 1845. 79.

KETCHUP

Hundreds of tomato ketchup recipes were published in America prior to the Civil War. While references to tomato ketchup have been located in the late eighteenth century and the early nineteenth, James Mease wrote the first known recipe for tomato ketchup; it was published in 1812. As it is not strained, this ketchup is thick. The second recipe, from the *American Farmer,* was published fifteen years later and was frequently copied. It is strained, and it makes a thin, juice-like, yellow ketchup.

Tomatoe, or Love-apple Catsup—James Mease, 1812

SLICE the apples thin, and over every layer sprinkle a little salt; cover them, and let them lie twenty-four hours; then beat them well, and simmer them half an hour in a bell-metal kettle; add mace and allspice. When cold, add two cloves of raw shallots cut small, and half a gill of brandy to each bottle, which must be corked tight, and kept in a cool place.

From James Mease, *Archives of Useful Knowledge; a Work Devoted to Commerce, Manufactures, Rural and Domestic Economy, Agriculture and the Useful Arts.* 2 vols. Philadelphia: David Hogan, 1812. Vol. 2, 306.

For Tomato Ketchup—half a Gallon—American Farmer, *1827*

As this is the season for making the best condiment for fish or steak that ever pantry was furnished with, I send the following recipe to the *American Farmer:*

Take—

a gallon of skinned tomatoes;

4 table spoonsful of salt;

4 do. black pepper;

half a spoonful alspice;

8 pods red pepper;

3 table spoonsful of mustard;

articles ground fine and simmered slowly in sharp vinegar, in a pewter basin, three or four hours, and then strained through a wire sieve and bottled close. It may be used in two weeks, but improves much by age. Those who like the article may add, after the simmering is over and the ingredients somewhat cooled, two table spoonsful of the juice of garlic. So much vinegar is to be used as to have half a gallon of liquor when the process is over. To my taste this is superior to any West India ketchup that I have ever met with, and it is withal *an excellent remedy for dyspepsia.* Cousin Tabitha.

From *American Farmer,* 9 (August 31, 1827): 191.

TOMATO PASTE

Today tomato paste and tomato purée are prepared commercially by evaporating the water from the pulp of the tomato after removing the seeds and skins. A purée consists of eleven to twenty-two percent solids, while paste consists of twenty-eight to forty-five percent solids. Although salt, some herbs and other natural herbs may be added, the Food and Drug Administration forbids the addition of other vegetable products such as carrots, apples, or beet roots to what is labeled tomato paste or purée.[1] Needless to say, these standards did not apply to the recipes below.

The *Cambridge Chronicle*'s tomato paste is unseasoned tomato pulp. It is closer to today's commercial definition of tomato paste. Sarah Rutledge's "Italian Tomato Paste" is more accurately a vegetable paste as it includes various vegetables and seasonings. Eliza Acton's recipe for "Purée of Tomatas" includes cream and flour and is like tomato sauce; it was intended for use on a variety of meat and poultry dishes.

Tomato Paste—Cambridge Chronicle, *1835*

Strain the juice of the Tomato through a fine cloth, and be careful that none of the seeds or rind pass with it. Evaporate the juice in the shade upon shallow plates or dishes—first, however, mixing salt (ad libitum) with the liquid. A paste is left, which will keep and preserve the true flavor of the fruit for several years. When intended to be used, a very small quantity is sufficient, the essence only being left after the evaporation.

From *Cambridge Chronicle,* September 26, 1835.

Italian Tomato Paste—Sarah Rutledge, 1847

Take a peck of tomatoes; break them and put them to boil with celery, four carrots, two onions, three table-spoonfuls of salt, six whole peppers, six cloves, and a stick of cinnamon; let them boil together (stirring all the time) until well done, and in a fit state to pass through a sieve; then boil the pulp until it becomes thick, skimming all the time. Then spread the jelly upon large plates or dishes, about half an inch thick; let it dry in the sun or oven. When quite dry, detach it from the dishes or plates, place it upon sheets of paper, and roll them up. In using the paste, dissolve it first in a little water or broth. Three inches square of the paste is enough to flavor two quarts of soup. Care should be taken to keep the rolls of paste where they will be preserved as much as possible from moisture.

From Sarah Rutledge, *The Carolina Housewife.* Charleston:
W. R. Babcock, 1847. 105.

Purée of Tomatas—Eliza Acton, 1845

Divide a dozen fine ripe tomatas, squeeze out the seeds, and take off the stalks; put them with one small mild onion (or more, if liked), and about half a pint of very good gravy, into a well-tinned stewpan or saucepan, and simmer them for nearly or quite an hour; a couple of bay-leaves, some cayenne, and as much salt as the dish may require should be added when they begin to boil. Press them through a sieve, heat them again, and stir to them a quarter-pint of good cream, previously mixed and boiled for five minutes with a teaspoonful of flour. This purée is to be served with calf's head, veal cutlets, boiled knuckle of veal, calf's brains, or beef palates. For pork, beef, geese, and other brown meats, the tomatas should be reduced to a proper consistency in rich and highly-flavoured brown gravy, or Spanish sauce.

From Eliza Acton, *Modern Cookery in All its Branches . . . Prepared for American Housekeepers. By Mrs. S. J. Hale.* From the 2d London edition. Philadelphia: Lea and Blanchard, 1845. 242.

TOMATO SAUCE

Dozens of tomato sauce recipes were published before the Civil War. Maria Eliza Rundell's "Tomata Sauce, for hot or cold Meats" and Richard Alsop's "Tomato, or Love Apple Sauce" were published in the United States in 1814. Rundell's recipe was borrowed from Alexander Hunter. Alsop's simple tomato sauce recipe was the first published by an American. Louis Eustache Ude's "Love-Apple Sauce" was complex and time-consuming to make. N. K. M. Lee offered four recipes for tomato sauce. Her Italian one contained the same ingredients as the French recipe with the addition of some seasonings. The sauce was as thick as a purée. Eliza Leslie's sauce was served with melted butter. *The Lady's Annual Register* added red wine and recommended using the sauce on meat, fish, and vegetables. During the 1840s hotel restaurants served tomato sauce on chicken, veal cutlets, mutton chops, sweetbreads, quails, and fish filets.

Tomata Sauce, for hot or cold Meats —Maria Eliza Rundell, 1814

Put tomatas, when perfectly ripe, into an earthen jar; and set it in an oven, when the bread is drawn, till they are quite soft; then separate the skin from the pulp; and mix this with capsicum vinegar, and a few cloves of garlic pounded, which must both be proportioned to the quantity of fruit. Add powdered ginger, and salt to your taste. (Some white-wine vinegar and Cayenne) may be used instead of capsicum vinegar. Keep the mixture in small wide-mouthed bottles, well corked, and in a dry, cool place.

From Maria Eliza Rundell, *A New System of Domestic Cookery.* New York: R. M'Dermut & D. D. Arden, 1814. 127.

Tomato, or Love Apple Sauce—Richard Alsop, 1814

Take the ripest and best tomatoes, carefully strip them of their outer peel, and cut out the insertion of the stalk and any spots which may be upon them, divide them into eight parts or slices, and take out a part, if not the whole, of the seeds. Put them into a saucepan, or even spider, with a very little butter previously melted, and cover them close, in order to keep in the steam; when nearly done, add to them salt and pepper to taste, and replacing the cover, let them stand a little time longer. Should it be desirable to thin the sauce, add a little water to it just before it is done.

From Richard Alsop, *The Universal Receipt Book or Complete Family Direction by a Society of Gentlemen in New York.* New York: I. Riley, 1814. 45.

Love-Apple Sauce—Louis Eustache Ude, 1828

Melt in a stew-pan a dozen or two of love-apples; (which before putting into the stew-pan, cut in two, and squeeze the juice and the seeds out,) then put two shalots, one onion, with a few bits of ham, a clove, a little thyme, a bay-leaf, a few leaves of mace, and when *melted,* rub them through a tammy. With this purée mix a few spoonsful of good *Espagnole,* a little salt and pepper. Boil it for twenty minutes, and serve up.[2]

From Louis Eustache Ude, *The French Cook.* Philadelphia: Carey, Lee and Carey, 1828. 9, 40.

Tomata Sauce, Italian—N. K. M. Lee, 1832

Take five or six onions, slice, and put them into a saucepan, with a little thyme, bay-leaf, twelve or fifteen tomatas, a bit of butter, salt, half a dozen allspice, a little India saffron, and a glass of stock; set them on the fire, taking care to stir it frequently, as it is apt to stick; when you perceive the sauce is tolerably thick, strain it like a *puree.*

From N. K. M. Lee, *The Cook's Own Book.* Boston: Munroe and Francis, 1832. 222.

Tomata Sauce—Eliza Leslie, 1832

Bake ten tomatas, with pepper and salt, till they become like a marmalade. Then add a little flour or grated bread crumbs, and a little broth or hot water. Stew it gently ten minutes, and before you send it to table add two ounces of butter and let it melt in the sauce.

From Eliza Leslie, *Domestic French Cookery, Chiefly Translated from Sulpice Barué.* Philadelphia: Carey & Hart, 1832. 21–22.

French Tomato Sauce—The Lady's Annual Register, *1843*

Put from fifteen to twenty Tomatos into a sauce pan with a little soup, some salt and pepper corns, make it boil until it is somewhat reduced—when it is of sufficient thickness rub it through a fine sieve—add to it if you please four or five spoonsful of red wine, at the moment of serving add a piece of butter of the size of an egg, which must be melted in the sauce. See that it is well seasoned. Serve it for particular dishes. It forms a very good relish to roasted pork.

From *The Lady's Annual Register, and Housewife's Almanac, for 1843.* Boston: T. H. Carter, 1843. 58.

TOMATO SOY

The second edition of Mary Randolph's *The Virginia House-wife,* published in 1825, featured "Tomata Soy," which was a misnomer, since it was not based on soybeans. It was, however, very popular. Other cookbook writers, such as Eliza Leslie, subsequently published similar recipes for tomato soy.

Tomata Soy—Eliza Leslie, 1837

For this purpose you must have the best and ripest tomatas, and they must be gathered on a dry day. Do not peel them, but merely cut them into slices. Having strewed some salt over the bottom of a tub, put in the tomatas in layers; sprinkling between each layer (which should be about two inches in thickness) a half pint of salt. Repeat this till you have put in eight quarts or one peck of tomatas. Cover the tub and let it set for three days. Then early in the morning, put the tomatas into a large porcelain kettle, and boil it slowly and steadily till ten at night, frequently mashing and stirring the tomatas. Then put it out to cool. Next morning strain and press it through a sieve, and when no more liquid will pass through, put it into a clean kettle with two ounces of cloves, one ounce of mace, two ounces of black pepper, and two tablespoonfuls of cayenne, all powdered.

Again let it boil slowly and steadily all day, and put it to cool in the evening in a large pan. Cover it, and let it set all night. Next day put it into small bottles, securing the corks by dipping them in melted rosin, and tying leathers over them.

If made exactly according to these directions, and slowly and thoroughly boiled, it will keep for years in a cool dry place, and may be used for many purposes when fresh tomatas are not to be had.

From Eliza Leslie, *Directions for Cookery.* Philadelphia: E. L. Carey & A. Hart, 1837. 224–25.

PRESERVING AND CANNING TOMATOES

BOTTLING

The first tomato recipe known to have been written by an American was Harriott Horry's "To Keep Tomatoos for Winter Use." Nicholas Appert's recipe provided the first commercial process for bottling tomatoes. Appert's experiments were considered the beginning of the commercial canning industry, although decades passed before canning became a significant business. Agricultural journals and cookbooks published numerous recipes for bottling tomatoes and also attempted to entertain their readers. The poem "To Preserve Tomatoes," first published in the *American Agriculturist* in 1849, combines entertainment with utility. It is the only known tomato poem that was published before the Civil War.

To Keep Tomatoos for Winter use —Harriott Pinckney Horry, 1770

Take ripe Tomatas, peel them, and cut them in four and put them into a stew pan, strew over them a great quantity of Pepper and Salt; cover it up close and let it stand an Hour, then put it on the fire and let it stew quick till the liquor is intirely boild away; then take them up and put it into pint Potts, and when cold pour melted butter over them about an inch thick. They commonly take a whole day to stew. Each pot will make two Soups.

N.B. if you do them before the month of Oct they will not keep.

From Harriott Pinckney Horry Papers, 28. From the collections of the South Carolina Historical Society.

Love Apples—Nicholas Appert, 1812

I gather love-apples very ripe, when they have acquired their beautiful colour. Having washed and drained them, I cut them into pieces, and dissolve them over the fire in a copper vessel well tinned. When they are well dissolved and reduced one third in compass, I strain them through a sieve sufficiently fine to hold the kernels. When the whole has passed through, I replace the decoction on the fire, and I condense it till there remains only one third of the first quantity. Then I let them become cool in stone pans, and put them in bottles, &c. in order to give them one, good boiling only, in the water-bath.

From Nicholas Appert, *The Art of Preservation*. New York: D. Longworth, 1812. 53–54.

To Preserve Tomatoes—American Agriculturist, *1849*

Six pounds of tomatoes first carefully wipe,
Not fluted nor green, but round, ruddy, and ripe;
After scalding, and peeling, and rinsing them nice—
With dext'rous fingers 'tis done in a trice—
Add *three* pounds of sugar, (Orleans will suit)
In layers alternate of sugar and fruit.
In a deep earthen dish, let them stand for a night,
Allowing the sugar and juice to unite!
Boil the sirup next day in a very clean kettle,
(Not iron, but copper, zinc, brass or bell-metal)
Which having well skimmed, 'till you think 'twil suffice
Throw in the tomatoes, first adding some spice—
Cloves, cinnamon, mace, or whate'er you like best—
'Twill add to the flavor, and give them a zest,
Boil slowly together until they begin
To shrink at the sides, and appear to fall in,
Then take them up lightly, and lay them to cool,
Still boiling the sirup, according to rule,
Until it is perfectly clear and translucent—
Your skill will direct you, or else there's no use in't—
Then into the jars, where the fruit is placed proper,
Pour boiling the sirup, direct from the copper.
After standing till cold, dip some paper in brandy,
Or rum, or in whisky, if that is more handy,
Lay it over the fruit with attention and care,
And run on mutton suet to keep out the air,
Then tie a strong paper well over the top—
And, "now that I think on't the story may stop."
If you'll follow these rules, your preserves, never fear,
Will keep in good order till this time next year.

From *American Agriculturist,* 8 (July 1849): 225.

CANNING

In America the commercial canning industry started during the 1840s. At the time, canning was a difficult and time-consuming task. A canning operation was usually run out of someone's barn. The recipe from the *Bangor Whig and Courier* encouraged farmers to can their own tomatoes. Despite its instructions to the cook to use a tin canister instead of a bottle, this recipe recommended Appert's water-bath method.

Recipe for Preserving Tomatos —**Bangor Whig and Courier,** *1849*

Pour boiling water on ripe tomatos and peel them. Cut them up and put them into two quart tin canisters, each having a hole two inches in diameter in the top. After they are filled have a circular piece soldered over the opening, leaving an awl-hole. Set the canisters into a kettle of boiling water for twenty minutes—stop the awl-hole with a spile of pine—finally when they are cool, cut off the spile even with the tin, and drop a drop of melted sealing-wax over it. On opening your canisters you will have tomatoes in perfection at any time of year.

N.B. Sprinkle a little salt and pepper on the fruit previous to packing in canisters.

From *Bangor Whig and Courier,* September 21, 1849.

Drying Tomatoes

Another way of preserving tomatoes was drying them in the sun or oven. This process became particularly popular the 1830s. Dr. James Dekay noted that the Turks dried them in the sun until they became a thick paste. The second recipe reproduced here for drying tomatoes is from Italy. After a cruise in the Mediterranean in 1848, Captain Engle of the U.S. Navy sent home Italian recipes for preparing and preserving tomatoes and saw the possibility of making a profit from the product.

Turkish Preparation of the Tomato by D[ekay] —**New York Farmer,** *1834*

I conceive we have nothing to equal it in giving pungency and flavor to our commonest dishes. It is a great desideratum to have it at all seasons of the year; and some of your readers will doubtless feel obliged by learning how to obtain it in a simple, easy, and economical manner. In Turkey, it is a universal favorite, and enters into the composition of all their sauces. I frequently saw it made, and the following recipe may be depended on, as it was corrected under the eyes of the good housewife herself.

The tomatoes are first washed in a weak brine, and hung up in a cool place to drain, until the following day; then squeeze them thoroughly by hand, throwing away the skins. The pulpy mass is strained through a fine cloth to prevent the seeds from passing through. It is then salted; put into shallow earthen plates, or dishes, and exposed to the sun for 12 days, or until it becomes a thick paste. It should be stirred with a wooden spoon, twice a day, while exposed to the sun. With respect to the quantity of salt to be added to the paste, the rule is, to put a hand-ful and a half to the pulp of a hundred tomatoes, if large, and less, if smaller.

Those who prepare tomatoes in this way will be surprised at the small quantity obtained, but their surprise will cease when they learn how far it will go. A bit not larger than a Lima bean will be sufficient to flavour the soup of a family of 20 persons; and a much smaller quantity for sauces. A small pot which I brought with me, containing about half a pint, lasted my family more than a year, and we used it very freely.

By stirring it frequently, fresh portions are exposed to the sun, and the salt is more thoroughly incorporated with it. The rule of 12 days holds good at Constantinople, and I should think would be sufficient here. At any rate, it should be thoroughly dried (covering it over at night) until it becomes of the consistence of hard butter.

From *New York Farmer,* 7 (November 1834): 323–24.

Spezia Method of Making "Conserve of Tomatoes" —Germantown Telegraph, *1850*

Take the Tomatoes well ripe, cut them in four parts, put them into a large sauce-pan to boil slowly, stirring them often so as to prevent them from sticking to the sides or bottom, which will cause them to burn; when they are well cooked, pass them through a sieve, saving carefully all which pass through, rejecting the seeds, skin, &c. Then place them to boil again in a clean sauce-pan, until they become thick; stirring well and frequently as before mentioned. When the substance becomes thick, take it out of the pan and spread it on plates of copper or tin, in the sun, to dry; and before it becomes quite dry, make it up in rolls about the size of a large sausage, rubbing the hands first with oil so as to prevent it from sticking to them. Then place the rolls in the sun to dry, turning them two or three times a day. Then place them in boxes, where they will keep for years.

Observe, that in moist or damp seasons, should any mould appear on them, you must rub your hands with oil and rub the rolls, when you can return them to the box.

From *Germantown Telegraph,* September 4, 1850.

Tomato Pickles

Pickling, another means of preserving tomatoes, was recommended in the United States as early as 1804. *The Lady's Book*'s recipe for "Ripe Tomato Pickles" exemplifies many pickling recipes.

Ripe Tomato Pickles—The Lady's Book, *1831*

Take ripe tomatos, and prick them with a fork or pointed stick, put them into any kind of vessel, salt each layer thickly; let them remain in the salt about eight days; at the expiration of eight days, put them for one night in a vessel of vinegar and water; then to a peck of tomatos and a bottle of good mustard, put half an ounce of cloves, half an ounce of pepper, and a dozen large onions sliced, pack them in a jar, placing a layer of onions and spices between layers of tomatos. In ten days the pickles will be in good eating order.

From *The Lady's Book,* 2 (March 1831): 168.

DESSERTS AND SWEETS

TOMATO FIGS

A Mrs. Steiger of Washington, D.C., gave a sample of her tomato figs to H. L. Ellsworth, the commissioner of the U.S. Patent Office. Ellsworth was so taken with the recipe that he sent it to the *American Farmer* with a letter extolling its virtues. He deeply regretted, he said, that "so many valuable improvements are lost to the world, barely for the want of publicity." This recipe received extensive publicity. It was among the few "original" recipes to be published in an agricultural journal and later to be reprinted in cookbooks. It was frequently republished in various publications, including *Scientific American* and *Godey's Lady's Book,* the latter of which published it three times.

Tomato Figs—American Farmer, *1841*

Take six pounds of sugar to one peck (or 16 lbs) of the fruit. Scald and remove the skin of the fruit in the usual way. Cook them over a fire, their own juice being sufficient without the addition of water, until the sugar penetrates and they are clarified. They are then taken out, spread on dishes, flattened and dried in the sun. A small quantity of the syrup should be occasionally sprinkled over them whilst drying; after which, pack them down in boxes, treating each layer with powdered sugar. The syrup is afterwards concentrated and bottled for use. They keep well from year to year, and retain surprisingly their flavor, which is nearly that of the best quality of fresh figs. The pear shaped or single tomatoes answer the purpose best. Ordinary brown sugar may be used, a large portion of which is retained in the syrup.

From *American Farmer,* 3d ser., 3 (August 18, 1841): 97.

TOMATO JELLY

In the early nineteenth century, the terms *jam, jelly,* and *preserves* were not well differentiated. Jelly was a semisweet, translucent condiment made by boiling fruit juice with sugar and pectin or gelatin. It was usually spread on toast or used as a dessert filling. Jam differed from jelly in that it was made from mashed fruit rather than the juice and was much thicker. Preserves contained pieces of the fruit. The two recipes below exemplify tomato jelly recipes published prior to 1860. They did not include gelatin or pectin.

Tomato Jelly—American Farmer, *1840*

The process of making them into jelly is simple and easy. When the tomatoes are ripe stew them thoroughly in a brass kettle. Their own juice is sufficient for this purpose. When well stewed, strain them through a coarse cloth, add an equal part of sugar, and then boil gently for a few hours, when it will be fit for the jars, to keep and use as is found convenient or necessary.

From *American Farmer,* 3d ser., 2 (October 7, 1840): 155.

Tomato Jelly—Sarah Rutledge, 1847

Fill a large jar with slices of the ripest and best tomatoes; lay a cloth over the jar, and over that put a piece of dough, to keep in the heat; place the jar in a large pot of water, and boil four or five hours, constantly; then strain the juice through a coarse hair sieve, and to every pint of juice put a pound of brown sugar, if you wish the jelly very sweet, or half that quantity if to eat with meat. Add the whites of eight eggs to every gallon of juice, skim it, and boil till nearly half evaporated; then put it in glasses, and keep them in the sun till sufficiently thick.

A very good jelly to eat with meat may be made by putting salt, pepper, and a little mace and nutmeg instead of sugar.

From Sarah Rutledge, *The Carolina Housewife.* Charleston: W. R. Babcock, 1847. 168.

Tomato Marmalade

Tomatoes were used to make marmalades and preserves. Mary Randolph published two recipes for tomato marmalade. Many others were published during the following thirty years.

Tomata Marmalade—Mary Randolph, 1824

Gather full grown tomatas while quite green, take out the stems and stew them till soft, rub them through a sieve, put the pulp on the fire seasoned highly with pepper, salt, and pounded cloves; add some garlic, and stew together till thick; it keeps well, and is excellent for seasoning gravies, &c.&c.

From Mary Randolph, *The Virginia House-wife.* Washington: Davis and Force, 1824. 201–2.

Tomata Sweet Marmalade—Mary Randolph, 1824

Prepare it in the same manner, mix some loaf sugar with the pulp, and stew until it is a stiff jelly.

From Mary Randolph, *The Virginia House-wife.* Washington: Davis and Force, 1824. 202.

TOMATO PIES AND TARTS

Tomatoes were used in pies by the 1820s, and this dish became a favorite of the southern table. The first known recipe for tomato pie was published by Lydia Child. The *Saturday Courier*'s recipe was a typical example of the tomato tart recipes that emerged during the 1840s.

Tomatoes Pie—Lydia Maria Child, 1836

Tomatoes make excellent pies. Skins taken off with scalding water, stewed twenty minutes or more, salted, prepared the sauce as rich squash pies, only an egg or two more.[3]

From Lydia Maria Child, *The Frugal Housewife.* 20th ed. Boston: American Stationers' Company, 1836. 114–15.

*Tomato Tart—*Saturday Courier, *1841*

Roll out your dough very thin, and place it on the plate in which you intend baking your tart, and slice your tomatoes very thin; spread them over the dough also very thinly, take about two table spoonfuls of brown sugar, and one of ground cinnamon bark, spread the two over the tomato, bake it well, and you have a delightful tart.

From *Saturday Courier,* as in the *American Farmer,* 3d ser., 3 (September 8, 1841): 126.

MISCELLANEOUS RECIPES

FOOD FOR THE SICK

Although there were thousands of advertisements for tomato pills and dozens of articles speculating as to their real contents, only three recipes for tomato medicine have been located. The first was Dr. A. J. Holcombe's recipe for his tomato pills, printed in the *Botanico-Medical Recorder.* As he added nothing to the tomatoes in his preparation, it was similar to recipes for dried tomatoes or tomato paste. Samuel Thomson's recipe called for molasses and "spirits." Both Holcombe's and Thomson's preparations concoctions required the pulp, not the juice. Catherine Esther Beecher's recipe for "Tomato Syrup" called for only the juice combined with sugar.

Tomato Pills—Botanico-Medical Recorder, *1838*

I mash the tomatoes and press them in the same manner that I would apples, to make cider—having the press fixed in such manner, with straw, as to prevent the seeds or rinds from running out with the juice. I then reduce the juice by evaporation, over a slow fire, to the consistency of honey in cold weather, or to that of stiff tar. In this state, if it be put into wide mouthed bottles, or small jars, and secured entirely from the air, it may be kept as long as necessary, perhaps for several years. This extract alone, taken in doses of 30 to 40 grains, will act as a mild cathartic; it has a fine effect upon the biliary organs, and is also diuretic and diaphoretic.

From various experiments on the fruit, in its different stages, I have proved to my own satisfaction, that the green fruit, when fully grown, is much better for medical uses, than the ripe. Try it, gentlemen, and you will be satisfied.

From *Botanico-Medical Recorder,* 6 (June 30, 1838): 317.

Tomato Pills—Samuel Thomson, 1841

Take one bushel of fresh gathered tomatoes, bruise and squeeze out the juice through a coarse cloth and let it stand for twelve hours, then pour off the juice from the sediment, and simmer it to the thickness of molasses; then take out what you wish for syrup, and simmer the remainder to the consistence of tar, and form into pills. Sweeten the syrup with molasses, and add sufficient spirits to keep from souring.

Dose—of the pills, from four to six at night, and varied at discretion, and of the syrup, from half to a wineglassful, three or four times a day.

From Samuel Thomson, *The Thomsonian Materia Medica, or Botanic Family Physician.* 13th ed. Albany, N.Y.: J. Munsell, 1841. 655–56.

Tomato Syrup—Catherine Esther Beecher, 1846

Express the juice of ripe tomatoes, and put a pound of sugar to each quart of the juice, put it in bottles, and set it aside. In a few weeks it will have the appearance and flavor of pure wine of the best kind, and mixed with water is a delightful beverage for the sick. No alcohol is needed to preserve it.

The medical properties of the tomato are in high repute, and it is supposed that this syrup retains all that is contained in the fruit.

From Catherine Esther Beecher, *Miss Beecher's Domestic Receipt Book.* New York: Harper & Brothers, 1846. 197.

TOMATO WINE

Recipes for tomato wine were published beginning in the 1840s. J. P. Nichols's recipe appeared in Chicago's *Prairie Farmer* and was basically tomato juice with sugar. Nichols considered it medicinal. Tomato wine recipes have been published regularly since the 1840s. After the Civil War these recipes became more complex.

Tomato Wine—Prairie Farmer, *1845*

To one quart of juice, put a pound of sugar, and clarify as for sweetmeats.

The above is very much improved by adding a small proportion of the juice of the common grape. The subscriber believes this wine far better and much safer for a tonic, or other medical uses, than the wines generally sold as Port Wines, &c. for such purposes. It is peculiarly adapted to some diseases and states of the system, and is particularly recommended for derangements of the liver. J. P. Nichols, Chicago.

From *Prairie Farmer,* 5 (July 1845): 168.

IMITATIONS

If a fruit or vegetable was out of season or otherwise unavailable, cooks substituted other products. One unusual culinary characteristic of the tomato was its great variability in color, shape, size, and acid content. This permitted it to be substituted for many products. Emma

Roberts used tomatoes to imitate the taste of a guava. The *Southern Planter* recipe substituted tomatoes for peaches. Eliza Leslie's recipe substituted tomatoes for honey.

A Preserve of Tomatos in Imitation of Guavas —Emma Roberts, 1844

Take the seeds out of unripe tomatos, and set them over a slow fire in weak sugar and water, until they are green. Then take out the tomatos, add sugar to the syrup, boiling it down until it is very strong, and of a good consistence. Pour it boiling over the tomatos, and let them remain in it until cold. Then repeat the process as often as appears necessary, but not sufficiently so to make them shrink. Should they be preserved ripe, pour the boiling syrup upon them, repeating it every two or three days until the sugar has completely penetrated the fruit. The addition of lemon-juice squeezed upon the tomatos, and a third or fourth part of strawberry-jam mixed with the syrup, will assist in the resemblance of the tomato to the guava, as will also a glass of portwine; but, as these would spoil the colour of green tomatos, they must only be put to those that are ripe.

From Maria Eliza Rundell, *A New System of Domestic Cookery,* ed. Emma Roberts. Philadelphia: Carey and Hart, 1844. 192.

Tomatoes Preserve —Southern Planter and Family Lyceum, *1832*

The tomato is favorably mentioned in your last number: it is a valuable vegetable. But I do not recollect, that in the variety of uses to which it has been applied, your paper assigns it any place among the different species of preserves. As we are deprived this season of that pride of the fruit of Georgia, *the peach,* it may be of service to housekeepers to know that the tomato forms a most admirable substitute for the peach, as a preserve. The flavor is almost precisely the same—it looks as well, and is altogether an excellent article for the tea table.

Directions—Take good ripe tomatoes, peel them and preserve them with good brown or loaf sugar. If not peeled they burst, and do not retain the consistency so much desired by housekeepers, though they are very good without peeling. I give you this, at this time, that the industry of the fair hands about your flourishing town, may profit by it, before *Jack Frost* shall cut off their hopes from this new source of table ornament and luxury. Yours, &c Sweet Tooth

P.S. Those who live in town will at all times find this a cheap substitute for peaches as the latter cost something. Remember "a penny saved is a penny gained."

From *Southern Planter,* as in the *American Farmer,* 14 (November 16, 1832): 286.

Tomata Honey—Eliza Leslie, 1838

To each pound of tomatas, allow the grated peel of a lemon and six fresh peach-leaves. Boil them slowly till they are all to pieces; then squeeze and strain them through a bag. To each pint of liquid allow a pound of loaf-sugar, and the juice of one lemon. Boil them together half an hour; or till they become a thick jelly. Then put it into glasses, and lay double tissue paper closely over the top. It will be scarcely distinguishable from real honey.

From Eliza Leslie, *Directions for Cookery.* 3d ed. Philadelphia: E. L. Carey & A. Hart, 1838. 441–42

NOTES

1. Peter G. Goose and Raymond Binsted, *Tomato Paste, Purée, Juice & Powder* (London: Food Trade Press, 1964), 2–5.

2. Ude's recipe for "Grand Espagnole" is

> Besides some slices of ham, put into a stew-pan some slices of veal. Moisten the same as for the *coulis;* sweat them in the like manner; let all the *glaze* go to the bottom, and when of a nice red colour, moisten with a few spoonsful of first *consommé,* to detach the *glaze:* then pour in the *coulis.* Let the whole boil for half an hour, that you may be enabled to remove all the fat. Strain it through a clean tammy. Remember always to put into your sauces some mushrooms, with a bunch of parsley and green onions.
>
> Louis Eustache Ude, *The French Cook* (Philadelphia: Carey, Lee and Carey, 1828), 9.

3. Child's recipe for "Pumpkin and Squash Pie" is as follows:

> For common family pumpkin pies, three eggs do very well to a quart of milk. Stew your pumpkin, and strain it through a sieve, or colander. Take out the seeds, and pare the pumpkin, or squash, before you stew it; but do not scrape the inside; the part nearest the seed is the

sweetest part of the squash. Stir in the stewed pumpkin, till it is as thick as you can stir it round rapidly and easily. If you want to make your pie richer, make it thinner, and add another egg. One egg to a quart of milk makes very decent pies. Sweeten it to your taste, with molasses or sugar; some pumpkins require more sweetening than others. Two tea-spoonfuls of salt; two great spoonfuls of sifted cinnamon; one great spoonful of ginger. Ginger will answer very well alone for spice, if you use enough of it. The outside of a lemon grated in is nice. The more eggs, the better the pie; some put an egg to a gill of milk. They should bake from forty to fifty minutes, and even ten minutes longer, if very deep.

Lydia Maria Child, *The Frugal Housewife* (Boston: American Stationers' Co., 1836), 66–77.

Part III

BIBLIOGRAPHY AND OTHER SOURCES

Painting by Paul Lacroix, *Still Life with Tomato,* 1865.

Bibliography and Other Sources

General Reference Works

Extensive references from various works have been cited in the Notes. For reasons of space, they are not repeated here. For those interested, an extensive bibliography on tomato references published or written before the Civil War is under preparation for publication.

Arber, Agnes. *Herbals, Their Origin and Evolution: A Chapter in the History of Botany 1470–1670.* Cambridge: University Press, 1938.

Bitting, Katherine. *Gastronomic Bibliography.* Reprint by Holland Press, London, 1981.

Burr, Fearing. *The Field and Garden Vegetables of America.* Reprint by American Botanist, Chillicothe, Ill. 1988.

Demaree, Albert Lowther. *The American Agricultural Press 1819–1890.* New York: Columbia University Press, 1941.

Hazlitt, W. C. *Old Cookery Books and Ancient Cuisine.* London: Elliot Stock, 1902.

Hedrick, Ulysses P. *History of Horticulture in America to 1860.* New York: Oxford University Press, 1950.

Henrey, Blanche. *British Botanical and Horticultural Literature before 1800.* 2 vols. London: Oxford University Press, 1975.

Hoornstra, Jean, and Trudy Jeath. eds. *American Periodicals 174–1900.* Ann Arbor, Mich.: University Microfilms International, 1979.

Hocker, Sally Haines. *Herbals and Closely Related Medico-botanical Works, 1472–1753.* Lawrence: University of Kansas Libraries, 1985.

Lowenstein, Eleanor. *Bibliography of American Cookery Books, 1742–1860.* New York: Antiquarian Society, 1972.

Romaine, Lawrence B. *A Guide to American Trade Catalogs 1744–1900.* New York: R. R. Bowker, 1960.

Rudolph, G. A. *Receipt Book and Household Manual.* Bibliography series, no. 4. Manhattan: Kansas State University, 1968.

Stafleu, Frans A., and Richard S. Cowan. *A Selective Guide to Botanical Publications and Collections with Dates, Commentaries and Types.* 2d ed. The Hague: Bohn, Scheltema & Holkena, 1981.

Stuntz, Stephen Conrad. *List of the Agricultural Periodicals of the United States and Canada Published during the Century July 1810 to July 1910.* U.S. Department of Agriculture Miscellaneous Publication no. 398. Washington, D.C.: U.S. Government Printing Office, 1941.

CULINARY HISTORY

The works listed below focus on general culinary history, the history of American food, and the history of specific food products. A few of these appeared in the Notes.

Cummings, Richard Osborn. *The American and His Food: A History of Food Habits in the United States.* Chicago: University of Chicago Press, 1940.

Fletcher, Stevenson Whitcomb. *The Strawberry in North America: History, Origin, Botany, and Breeding.* New York: Macmillan, 1917.

Foust, Clifford M. *Rhubarb, the Wondrous Drug.* Princeton, N.J.: Princeton University Press, 1992.

Fussell, Betty. *The Story of Corn.* New York: Knopf, 1992.

Furnas, C. C., and S. M. Furnas. *The Story of Man and his Food.* New York: New Home Library, 1941.

Gannett, Lewis. *Cream Hill: Discoveries of a Weekend Countryman.* New York: Viking, 1949.

Hale, William Harlan. *The Horizon Cookbook and Illustrated History of Eating and Drinking through the Ages.* 2 vols. New York: American Heritage, 1968.

Hess, John, and Karen Hess. *The Taste of America.* New York: Grossman, 1977.

Hess, Karen. *The Carolina Rice Kitchen: The African Connection.* Columbia: University of South Carolina Press, 1992.

Hobhouse, Henry. *Seeds of Change: Five Plants that Transformed Mankind.* New York: Harper & Row, 1986.

Hooker, Richard J. *The Book of Chowder.* Boston: Harvard Common Press, 1978.

Hooker. *A History Food and Drink in America.* Indianapolis & New York: Bobbs-Merrill, 1981.

Jones, Evan. *American Food: The Gastronomic Story.* 2d ed. New York: Vintage Books, 1981.

Kahn, E. J. *The Staffs of Life.* Boston: Little, Brown, 1985.

Kreidberg, Marjorie. *Food on the Frontier: Minnesota Cooking from 1850 to 1900 with Selected Recipes.* Saint Paul: Minnesota Historical Society Press, 1975.

Mariani, John F. *The Dictionary of American Food and Drink.* New York: Ticknor and Fields, 1983.

Mintz, Sidney W. *Sweetness and Power: The Place of Sugar in Modern History.* New York: Viking, 1985.

Multhauf, Robert P. *Neptune's Gift: A History of Common Salt.* Baltimore: Johns Hopkins University Press, 1978.

Revel, Jean-François. *Culture and Cuisine, A Journey Through the History of Food.* Garden City, N.Y.: Doubleday, 1982.

Ritchie, Carson I. A. *Food in Civilization: How History Has Been Affected by Human Tastes.* New York: Beaufort Books, 1981.

Root, Waverly, and Richard de Rochemont. *Eating in America; A History.* New York: Echo, 1981.

Root, Waverly. *Food: An Authoritative and Visual History and Dictionary of the Foods of the World.* New York: Simon and Schuster, 1980.

Salaman, Radcliff. *The History and Social Influence of the Potato.* Rev. ed. with a new introduction by J. G. Hawkes. Cambridge: Cambridge University Press, 1986.

Smith, R. E. F., and David Christian. *Bread & Salt, A Social and Economic History of Food and Drink in Russia.* Cambridge: Cambridge University Press, 1984.

Sokolov, Raymond. *Why We Eat What We Eat: How Columbus Changed the Way the World Eats.* New York: Simon & Schuster, 1993.

Tannahill, Reay. *Food in History.* Rev. ed. New York: Crown, 1988.

Tousaint-Samat, Maguelonne. [Anthea Bell, trans.] *History of Food.* Cambridge, Mass.: Blackwell, 1992.

Wilson, C. Anne. *Food and Drink in Britain.* Chicago: Chicago Academy Publishers, 1991.

Reprinted Cookery Books and Manuscripts

As original cookbooks and manuscripts are often difficult to locate, a number of recent reprints are listed below that may be available in libraries. Only reprints that are related to the historical period covered in this book or that are mentioned in the text have been included here.

Appert, Nicholas. *The Art of Preserving.* Reprint by the Mallinckrodt Collection of Food Classics.

Bryan, Lettice. *The Kentucky Housewife.* Reprint by the University of South Carolina Press, Columbia, 1991.

Chadwick, J. *Home Cookery.* Reprint by Oxmoor House, Birmingham, Ala., 1984.

Confederate Receipt Book; A Compilation of over One Hundred Receipts, Adapted to the Times. Reprint by the University of Georgia Press, Athens, 1960.

Child, Lydia Maria. [Allice M. Geffen, ed.] *The American Frugal Housewife.* Reprint by Harper & Row, New York, 1972.

Farmer, Fannie Merritt. *Boston Cooking-school Cook Book.* Reprint by Weathervane Books, New York, 1986.

Gowan, Judy, and Hugh Gowan. *Blue and Grey Cookery: Authentic Recipes from The Civil War Years.* Reprint by Daisy Publications, New Market, Md., N.d.

Haskell, Mrs. E. F. *Civil War Cooking: The Housekeeper's Encyclopedia.* Reprint by R. L. Shep, Mendocino, Calif., 1992.

Hess, Karen, ed. *Martha Washington's Booke of Cookery.* New York: Columbia University Press, 1981.

Hooker, Richard J., ed. *A Colonial Plantation Cookbook: The Receipt Book of Harriott Pinckney Horry, 1770.* Columbia: University of South Carolina Press, 1984.

The Housekeeper's Book. Reprint by the New Hampshire Publishing Co., Somersworth, 1972.

Josephson, Bertha E., ed. "Ohio Recipe Book of the 1820s," *Mississippi Valley Historical Review,* 36 (June 1949): 97–107.

Lee, N. K. M. *The Cook's Own Book.* Reprint by Arno Press, New York Times Books, 1972.

Leslie, Eliza. *Directions for Cookery.* Reprint by Arno Press, New York Times Books, 1973.

McDougall, Francis Harriet. *The Housekeeper's Book.* Reprint by the New Hampshire Publishing Co., Somersworth, 1972.

Randolph, Mary. *The Virginia House-wife.* Facsimile edition with historical notes and commentaries by Karen Hess. Reprint by the University of South Carolina Press, Columbia, 1984.

Rutledge, Sarah. *The Carolina Housewife or House and Home.* Facsimile edition with introduction and preliminary checklist of South Carolina cookbooks published before 1935 by Anna Wells Rutledge. Reprint by University of South Carolina Press, Columbia, 1979.

Simmons, Amelia. *American Cookery.* Reprint by William R. Erdmans, Grand Rapids, Mich., 1965.

Thornton, Phineas. *The Southern Gardener and Receipt Book.* Reprint by Oxmoor House, Birmingham, Ala., 1984.

Ude, Louis Eustache. *The French Cook.* Reprint by Arco, New York, 1978.

Weaver, William Woys. ed. *Sauerkraut Yankees: Pennsylvania German Food and Foodways.* Philadelphia: University of Pennsylvania Press, 1983. Translation of Gustav Sigismund Peters, *Die geschickte Hausfrau.*

Weaver, ed. *A Quaker Woman's Cookbook; The Domestic Cookery of Elizabeth Ellicott Lea.* Reprint by University of Pennsylvania Press, 1982.

Webster, A. L. *The Improved Housewife, or Book of Receipts.* Reprint by Arno Press, New York Times Books, 1973.

WORKS ON TOMATO HISTORY

Compared with works on other fruits and vegetables, relatively few have been published about tomato history. Those articles or other works that do provide useful information on the history of the tomato are listed below.

Behr, Edward. "A Flavorful Tomato, Revisited," *The Art of Eating,* 9 (Winter 1989): 1, 7.

Corbett, Wilfred. "The History of the Tomato," in *Fifteenth Annual Report of the Experimental and Research Station.* London: Cheshunt, 1930. Pp. 82–83.

Dillon, Clarissa. "Tomato Mania," *Living History,* 1 (Summer 1991): 1–8.

Fawcett, W. Peyton. "Happiness is a Ripe Love Apple," *Field Museum of Natural History Bulletin,* 41 (August 1970): 2–5.

Gannett, Lewis. *Cream Hill: Discoveries of a Weekend Countryman.* New York: Viking, 1949.

Grewe, Rudolf. "The Arrival of the Tomato in Spain and Italy: Early Recipes," *The Journal of Gastronomy,* 3 (Summer 1987): 67–83.

Hedrick, Ulysis P., ed. *Sturtevant's Edible Plants of the World.* New York: Dover, 1972. Facsimile of the 1919 edition originally titled *Sturtevant's Notes on Edible Plants.*

Jenkins, J. A. "The Origin of the Cultivated Tomato," *Economic Botany,* 2 (October-December 1948): 379–92.

Luckwill, Leonard C. *The Genus Lycopersicon: An Historical, Biological and Taxonomic Survey of the Wild and Cultivated Tomatoes.* Aberdeen University Studies, no. 120. Aberdeen: University Press, 1943.

Luckwill. "The Evolution of the Cultivated Tomato," *Journal of the Royal Horticultural Society,* 68 (1943): 19–25.

McCue, George A. "The History of the Use of the Tomato: An Annotated Bibliography," *Annals of the Missouri Botanical Garden,* 39 (November 1952): 289–348.

Moore, John Adam. "The Early History of the Tomato or Love Apple," *Missouri Botanical Garden Bulletin,* 23 (October 1935): 134–38.

Muller, Cornelius. *A Revision of the Genus Lycopersicon.* Washington, D.C.: United States Department of Agriculture, 1940. Miscellaneous Publication, No. 382.

Muller, Cornelius. "The Taxonomy and Distribution of the Genus Lycopersicon," *National Horticultural Magazine,* 19 (July 1940): 157–60.

Nimtz, Sharon, ed. "Tomato Imperative!" *Small Bites,* 1 (Summer 1991; revised Summer 1992): 1–8.

Richman, Irwin. "The History of the Tomato in America," *Proceedings of the New Jersey Historical Society,* 80 (July 1962): 151–73.

Sim, Mary B. *Commercial Canning in New Jersey, History and Early Development.* Trenton: New Jersey Agricultural Society, 1951.

Smith, Andrew F. "The Making of the Legend of Robert Gibbon Johnson and the Tomato," *New Jersey History,* 108 (Fall-Winter 1990): 59–74.

Smith. "Tomato Pills will Cure All Your Ills," *Pharmacy in History,* 33 (1991): 169–80.

Smith. "The History of Home-made Anglo-American Tomato Ketchup," *Petits Propos Culinaires,* 39 (December 1991): 35–45.

Smith. "Dr. John Cook Bennett's Tomato Campaign," *Old Northwest,* 16 (Spring 1992): 61–75.

Smith. "Authentic Fried Green Tomatoes?" *Food History News,* 4 (Summer 1992): 1–2.

Smith. "The Great Tomato Pill War of the Late 1830s," *Connecticut Historical Society Bulletin,* 56 (Winter-Spring 1991): 91–107.

Smith. "The Amazing Archibald Miles and his Miracle Pills: Dr. Miles' Compound Extract of Tomato," *Queen City Heritage,* 50 (Summer 1992): 36–48.

Weber, George F. "A Brief History of Tomato Production in Florida," *Proceedings of the Florida Academy of Sciences,* 4 (1940): 167–74.

TOMATO COOKBOOKS AND PAMPHLETS

Several works with large sections of tomato recipes have been published since the late nineteenth century. The first cookbook solely on tomatoes was published in 1968, and since then several such books or pamphlets have been published. Several new tomato cookbooks are under development.

Bailey, Lee. *Tomatoes.* New York: Clarkson Potter, 1992.
Bay Books. *The Amazing Tomato Cookbook.* Kensington, New South Wales: Bay Books, n.d.
Bevona, Don. *The Love Apple Cookbook.* New York: Funk and Wagnalls, 1968.
Carver, George Washington. "How to Grow the Tomato and 115 Ways to Prepare It for the Table," Bulletin no. 36, Tuskegee Institute Experiment Station, Tuskegee, Ala., 1936.
Dribin, Lois, Denise Marina, and Susan Ivankovich. *Cooking with Sun-Dried Tomatoes.* Tucson, Ariz.: Fisher Books, 1990.
DuBose, Fred. *The Total Tomato; America's Backyard Experts Reveal the Pleasures of Growing Tomatoes at Home.* New York: Harper Colophon, 1985.
Garden Way Publishing. *Tomatoes! 365 Healthy Recipes for Year-Round Enjoyment.* Pownal, Vt.: Storey Communications, 1991.
Hendrickson, Robert. *The Great American Tomato Book: the One Complete Guide to Growing and Using Tomatoes Everywhere.* Garden City, N.Y.: Doubleday, 1977.
Hobson, Phyllis. *Great Green Tomato Recipes.* Bulletin A-24, Pownal, Vt.: Storey Communications, 1978.
Livingston, A. W. *Livingston and the Tomato.* Columbus, Ohio: A. W. Livingston's Sons, 1893.
Michaelson, Mike. *The Great Tomato Cookbook.* Chicago: Greatlakes Living Press, 1975.
Nimtz, Sharon, and Ruth Cousineau. *Tomato Imperative!* New York: Little, Brown, 1994.
Old-Fashioned Tomato Recipes, including Green Tomato Recipes. Nashville, Ind.: Bear Wallow Books, 1981.
Simmons, Paula. *The Green Tomato Cookbook.* Seattle: Pacific Search, 1975.
Turman, Marianne, and Ray Turman. *Cooking with Fresh Tomatoes.* Sun City Center, Fla.: Printed for the Author, 1992.

TOMATO CULTURE

Carver, George Washington. "How to Grow the Tomato and 115 Ways to Prepare It for the Table," Bulletin no. 36, Tuskegee Institute Experiment Station, Tuskegee, Ala., 1936.
Day, J. W., D. Cummins, and A. I. Root. *Tomato Culture in Three Parts.* Medina, Ohio: A. I. Root Company, 1906.
Foster, Catherine. *Terrific Tomatoes.* Emmaus, Pa.: Rodale, 1975.
Krech, Inez M. *Tomatoes.* New York: Crown, 1981.

Livingston, A. W. *Livingston and the Tomato.* Columbus, Ohio: A. W. Livingston's Sons, 1893.

National Gardening Association. *Tomatoes; Growing, Cooking, Preserving.* New York: Villard, 1987.

Page, John. *Grow the Best Tomatoes.* Bulletin A-27. Pownal, Vt.: Storey Communications, 1979.

Pellett, Frank C., and Melvin A. Pellett. *Practical Tomato Culture.* New York: A. T. De La Mare, 1930.

Raymond, Dick and Jan. *The Gardens for All Book of Tomatoes.* Burlington, Vt.: National Association for Gardening, 1983.

Rundell, Mary G. *Texas Gardener's Guide to Growing Tomatoes.* Waco, Tex.: Suntex Communications, 1984.

Rupp, Rebecca. *Blue Corn & Square Tomatoes: Unusual Facts about Common Garden Vegetables.* Pownal, Vt.: Garden Way, 1987.

Tarr, Yvonne Young. *The Tomato Book.* New York: Vintage, 1976.

Tracy, William W. *The Tomato Culture.* New York: Orange Judd, 1907.

Watterson, John C. *Tomato Diseases: A Practical Guide for Seedsmen, Growers & Agricultural Advisors.* Saticoy, Calif.: Petroseed Co., 1985.

Wittwer, S. H., and S. Honma. *Greenhouse Tomatoes, Lettuce and Cucumbers.* East Lansing: Michigan State University Press, 1979.

Work, Paul. *The Tomato.* New York: Orange Judd, 1942.

Heirloom Seed Sources, Tomato Organizations, and Other Resources

Tomato organizations and clubs, along with newsletters and other publications, have been around since the beginning of the twentieth century. Listed below are organizations, projects, contests, and businesses that have resources related to different aspects of the tomato.

California Tomato Grower's Association
2529 West March Lane
Stockton, CA 95207
(209) 478–1761

Florida Tomato Committee
P.O. 140635
Orlando, FL 32814
(407) 894–3071

The Greater New Jersey Championship Tomato Weigh-In
P.O. Box 123
Monmouth Beach, NJ 07750
(908) 229–2395

The Heirloom Seed Project
The Landis Valley Museum
2451 Kissel Hill Road
Lancaster, PA 17601
(717) 569–0401

Seed Savers Exchange
3076 North Winn Road
Decorah, IA 52101
(319) 382-5990

Southern Exposure Seed Exchange
P.O. Box 158
North Garden, VA 22959
(804) 973-4703

The Tomato Club
114 East Main St.
Bogota, NJ 07603
(201) 488–2231

The Tomato Genetics Resource Center
Genetic Resources Conservation Program
University of California
Davis, CA 95616
(916) 757–8920

Tomato Growers Supply Company
P.O. Box 2237
Fort Myers, FL 33902
(813) 768–1119

The Tomato Seed Company
P.O. Box 1400
Tryon, NC 28782

Totally Tomatoes
Box 1626
Augusta, GA 30903

Index of Names

General Index